DR KAWASHIMA

TRAIN YOUR BRAIN MORE

60 Days to a Better Brain

PENGUIN BOOKS

Published by the Penguin Group
Penguin Books Ltd, 80 Strand, London WC2R 0RL, England
Penguin Group (USA) Inc., 375 Hudson Street, New York, New York 10014, USA
Penguin Group (Canada), 90 Eglinton Avenue East, Suite 700, Toronto, Ontario, Canada M4P 2Y3
(a division of Pearson Penguin Canada Inc.)
Penguin Ireland, 25 St Stephen's Green, Dublin 2, Ireland (a division of Penguin Books Ltd)
Penguin Group (Australia), 250 Camberwell Road, Camberwell, Victoria 3124, Australia
(a division of Pearson Australia Group Pty Ltd)
Penguin Books India Pvt Ltd, 11 Community Centre, Panchsheel Park, New Delhi – 110 017, India
Penguin Group (NZ), 67 Apollo Drive, Rosedale, North Shore 0632, New Zealand
(a division of Pearson New Zealand Ltd)
Penguin Books (South Africa) (Pty) Ltd, 24 Sturdee Avenue,
Rosebank, Johannesburg 2196, South Africa

Penguin Books Ltd, Registered Offices: 80 Strand, London WC2R 0RL, England

www.penguin.com

First published 2008
1

Copyright © Ryuta Kawashima/KUMON PUBLISHING Co., Ltd, 2004
First published in Japan in 2005 under the title 'NOU WO KITAERU OTONA NO
KEISAN DORIRU 2' by Kumon Publishing Co., Ltd
English translation rights arranged with Kumon Publishing Co., Ltd through
Japan Foreign Rights Centre
All rights reserved

The moral right of the author has been asserted

Printed and bound in Italy by Graphicom, srl

ISBN: 978–0–141–03550–5

www.greenpenguin.co.uk

Penguin Books is committed to a sustainable future
for our business, our readers and our planet.
The book in your hands is made from paper
certified by the Forest Stewardship Council.

Dear reader,

Early on in my career as a brain researcher, I had an idea which I was sure would bring me tons of funding for my work. I decided to use brain imaging devices to investigate the best way to stimulate our brains.

After twenty years of research, I believe I've found the answer. The best way to stimulate the brain is to perform simple mathematical calculations quickly, and to read books aloud. Surprising? Possibly. But as you will see in the coming pages, these two activities light up the brain like a switchboard.

With some added research, I found that these two activities can even help people retain mental clarity and stave off the mental effects of aging. I used these results to create a brain health workbook called *Train Your Brain*, hoping that I could help make brain health an important social priority.

In 2003, *Train Your Brain* was published in Japan. To everyone's shock, the book and its sequel *Train Your Brain More* became hits, eventually selling millions of copies! As a result of the books, I constantly receive warm messages from people who tell me "your books made exercising my brain part of my daily activity" and "your therapy has improved my memory and made me more energetic". This really delights me because I feel it is a wonderful thing to be able to help others through one's work.

Now, I am very pleased to be able to make a new version of *Train Your Brain More* available to people outside of Japan. I believe many readers around the world will find this book extremely beneficial.

It's important to realize that, just as you exercise your body, you must regularly stimulate and exercise your brain. This is the secret to remaining young mentally. So, if you want to stay young at heart, and sharp mentally, take a few minutes a day to *Train Your Brain More!*

Sincerely,   Prof. Ryuta Kawashima MD

Table of Contents

Introduction

Ryuta Kawashima
M.D., Tohoku University

Why should you train your brain?

This book is a follow-up to *Train Your Brain: 60 Days to a Better Brain*. I've received lots of letters from readers since the first book came out, and I'm delighted to know that people all over the world are now making these drills a regular part of their daily routine. To keep things fresh, we have made one change to the format of the exercises. We have included some simple division problems for you to solve as you work through your daily brain-training exercises. Brain function naturally begins to deteriorate after our twenties, just like our physical and muscular strength gradually weaken as we age. However, just as you can maintain your physical strength if you exercise regularly, you can keep your brain power from deteriorating by providing daily stimulation for your brain.

In my neuroscience lab, I developed the exercises in this book in order to activate the largest regions of the brain. These brain health exercises increase the delivery of oxygen, blood, and various amino acids to the prefrontal cortex. The result is more neurons and neural connections, which are characteristics of a healthy brain.

Who is this book for?

• **Adults with the following symptoms:**

> ★Increasing forgetfulness
> ★Difficulty remembering people's names, spelling words, or expressing thoughts.

• **Adults who wish to work on the following:**

> ★Creativity
> ★Memory skills
> ★Communication skills
> ★Slowing the mental effects of aging

How can you keep your brain healthy?

In order to maintain your physical health, you have to (1) exercise regularly, (2) eat healthily and (3) sleep well. In like manner, in order to keep your brain healthy, you need to (1) exercise your brain regularly, (2) eat healthily and (3) sleep well. As an adult, you are responsible for your own diet and sleep on a daily basis. This book is exclusively designed to help you get used to also training your brain every day.

Simple calculations really work!

While browsing through this workbook, you may have noticed it consists only of simple calculations. Some of you may be wondering why an adult, who is intellectually active at work and at home, would need to do such elementary school-level math. Through my research I found that simple calculations activate the brain more effectively than any other activity. I also discovered that the best way to activate the largest regions of the brain was to solve these calculations quickly. That is why I have created the easy-to-solve problems you see in this workbook to help you Train Your Brain!

Calculation and oral reading exercises are the optimal training methods according to the latest brain research

My latest research proved that reading aloud, solving simple calculations, and writing activate the brain most effectively.

The pictures on the right show images of the brain taken by a brain-imaging device during various activities (see note 1 on page 12). The red color on the images indicates an increase in regional cerebral blood flow, while the yellow color indicates a larger increase in blood flow to areas where brain activity is most vigorous.

Let's take a look at the images in Charts B and C. Notice that while you are solving calculation problems quickly, there are several red areas in both hemispheres of the brain. Chart B indicates that the visual area at the posterior part of the brain is actively working. Other active parts are the inferior temporal gyrus, which detects the shapes of numbers; Wernicke's area, which recognizes the meanings of words; the angular gyrus, which is the area for calculating; and most importantly, the prefrontal cortex, which is the area for thinking and learning. In comparison, Chart C shows your brain while you are solving simple calculations slowly. The same parts of the brain are working, but the active areas are smaller in size. Solving difficult calculations, surprisingly, does not activate much of your prefrontal cortex at all (see Chart D).

While you are thinking and watching TV, (see charts A and E) most of the different parts of the brain are barely functioning. These charts show that solving simple calculations quickly is the most effective way of activating your brain.

A While you are thinking

left hemisphere   right hemisphere

anterior   posterior   anterior

This is the state of your brain while you are deep in thought. Note the tiny active part in the left hemisphere of the prefrontal cortex (see note 2 on page 12).

E While you are watching TV

left hemisphere   right hemisphere

This is the state of your brain while you are watching TV. The only active parts in either hemisphere of the brain are the visual-oriented occipital lobe and the audio-oriented temporal lobe (See note 2).

B While you are solving simple calculation problems quickly

left hemisphere   right hemisphere

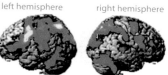

This is the state of your brain while you are quickly solving simple calculations of the kind contained in this book. As you can see, multiple parts of both hemispheres of the brain are actively working. Note how active the prefrontal cortex is.

C While you are solving simple calculations slowly

left hemisphere   right hemisphere

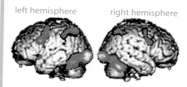

This is the state of your brain while you are slowly solving simple calculations of the kind contained in this book. You can see that the brain is much more active when you solve calculations quickly.

D While you are solving difficult calculations

left hemisphere   right hemisphere

This is the state of your brain while you are solving difficult calculations. A part of the prefrontal cortex and a part of the left hemisphere of the brain are activated. Note the inactive state of the right hemisphere of the brain.

F While you are writing

left hemisphere   right hemisphere

This is the state of your brain while you are writing. Note the vigorous state of the prefrontal cortex in both hemispheres of the brain.

G While you are silently reading

left hemisphere   right hemisphere

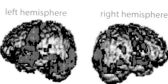

This is the state of your brain while you are silently reading. Many different regions in both hemispheres of the brain are activated.

H While you are reading aloud

left hemisphere   right hemisphere

This is the state of your brain while you are reading aloud. Active areas in both hemispheres of the brain are much larger than in Chart G. Research has also shown that the brain becomes more active when you read faster.

Brain training boosted the ability to remember by twenty percent

My research team carried out research with elementary school students. We counted How many words they could memorize within two minutes and found that, on average, they could memorize 8.3 words (the equivalent figure for adults is 12.2). When we conducted the same test after a two-minute calculation exercise, the average word count remember increased to 9.8 and after two minutes of reading out loud the average increased to 10.1. Our results show that these exercises boosted the children's ability to remember by more than twenty percent.

The calculation and reading aloud exercises acted as a warm-up for the students, allowing them to perform better on the word memorization tests.

Simple calculation and oral reading exercises alleviated symptoms of dementia

My team also conducted an experiment with twelve dementia Alzheimer type patients. We gave our patients a ten-minutes-a-day writing and oral reading exercise and a ten-minutes-a-day calculation exercise to be performed two to five days a week. Cognitive and prefrontal cortex function of non-participant subjects that did not do the calculation and oral reading exercises deteriorated during the six-month follow up. However, with participant subjects that did do the exercises, we succeeded in preventing deterioration of their cognitive function, as well as improving their prefrontal cortex function.

This is an exceptional achievement on a global scale - rarely has the deterioration of the cognitive function of dementia Alzheimer type patients been slowed or diminished.

1 Change in word memorization

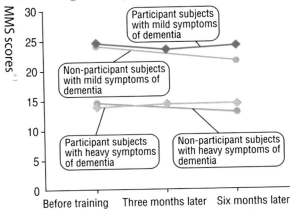

2 Change in cognitive function

* 1 MMS tests evaluate your cognitive abilities such as comprehension and judgment.

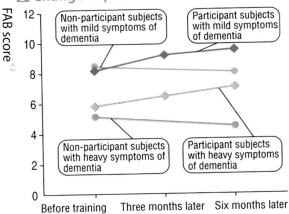

3 Change in prefrontal cortex function

* 2 FAB tests evaluate your prefrontal cortex function by testing your abilities to communicate and control behavior.

How to train your brain using this book

1 First, check how well your brain is functioning currently

Check the status of your brain function by using the Pre-training Prefrontal Cortex Evaluation. Please see instruction **5** to learn more about evaluation procedures.

2 Train your brain a few minutes every day

The most crucial point in any kind of training is continuity. It is ideal to train your brain in the morning, when your brain is the most active. It is also important to eat before you do the calculation exercise; otherwise, the effects will be reduced by half.

Many of you may have noticed it takes you more time to calculate in the afternoon and at night than in the morning. That is because the brain functions more effectively in the morning than at any other time of the day. I suggest you carry out the exercises at the same hour every day, if possible.

3 Tips for the program

Please solve the calculation problems on the front and back of one worksheet each day. Try to solve the problems quickly, but do not worry if you make mistakes or are slow at first. All evaluations are only for the purposes of tracking your own progress.

Worksheet target time

• **One minute :**
Gold Medal. If you often calculate manually, or regularly use math at work, you may be able to achieve this target. In other words, you can call yourself a Numbers Whiz if you reach this level.

• **One and a half minutes :**
Silver Medal. With continued effort, you can reach this level. If you do achieve this goal, your calculation abilities must be higher than average. You may call yourself a Calculation Expert.

• **Two minutes :**
Bronze Medal. You will be able to attain this level if you set your mind to it - please keep working toward this goal. If you do attain this level, you should be known as a Calculation Master.

The Division Problems

A lot of the time, you will find that the answers to the division problems are not whole numbers. When this happens, write the nearest whole number in the answer box, and then fill in the amount left over in the box marked "Remainder". For example: 14÷3= 4 Remainder 2 Always remember to make sure that the remainder is smaller than the number you are dividing by!

4 Stop and evaluate after every fifth day

This 60-Day Calculation Exercise Program is comprised of calculation worksheets to be solved one per day from Monday through Friday and Prefrontal Cortex Evaluations to be taken during weekends. If you do not wish to stop training during weekends, or if your busy schedule does not allow you to study every weekday as I recommend, then take the Prefrontal Cortex Evaluation after every five study days.

After completing each Evaluation, please record your results in My Brain Training Chart at the end of the appendix. You may clearly recognize positive changes in your brain as you log the results each week (see note 3 on page 12). Please try to exercise continuously; if there is an interval between exercises, the effects may become less noticeable.

5 How to evaluate your prefrontal cortex

Before starting your calculation program, please take the Pre-Training Prefrontal Cortex Evaluation. After you start your training, please evaluate your prefrontal cortex after every five study days using the Prefrontal Cortex Evaluations at the end of the book. It may be a good idea to have a member of your family or a friend measure the time required for you to solve the problems.

Counting Test

Measure the time required for you to count the numbers from 1 to 120 aloud as fast as you can. You must pronounce every number articulately. This test evaluates the general function of the prefrontal cortex in both the left and right hemispheres. Research has shown that the results of counting tests like this one are closely related to the mathematical abilities of each person. Please set your own goal and try to achieve your personal target time.

Word Memorization Test

Each word list consists of 30 simple words. Please memorize as many words as you can within two minutes. After two minutes, turn the paper over and write down as many words as you can remember. The number of correct answers will be your score. The aim of this test is to evaluate your prefrontal cortex activity in the left hemisphere, which is linked to short-term memory.

Stroop Test

At the end of the book, you will see a chart of color names. Often the color names are printed in a color different from the color they actually label. The test requires you to *say the color the word is printed in aloud*. In other words, if you see the word "Blue" printed in red ink, you should say "Red," not "Blue." Try to ignore the words that you see, and report only the color of the ink.

Please get used to the procedure by first using the chart at the top of each Stroop Test. Then proceed to the test. Measure and record the time required for you to name all of the colors. The aim of this test is to evaluate the general function of your prefrontal cortex area in both the left and right hemispheres. There are no target times or scores, as the results of the test vary greatly according to individuals. Please just try to do better than your previous week's results.

■ **Example**

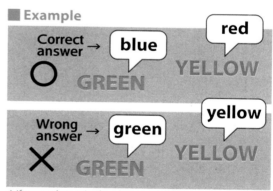

* If you make a mistake, please go back to where you made the mistake and start over.

6 When you're finished with this workbook...

Now that calculation has become a part of your daily routine, it is important to continue your calculation exercise every day. Once you stop training your brain, your brain function will begin to decline again slowly. Please go back to Day 1 of the book and repeat the training.

• **Note 1 :**

Brain-imaging devices create three-dimensional images of human brain function without harming the body or the brain. Functional magnetic resonance imaging (fMRI) and near-infrared spectroscopy are the two major technologies used in brain science studies at this time.

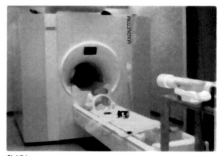

fMRI apparatus

• **Note 2 :**

There are four parts in both hemispheres of the human brain which control individual function: the frontal lobe (motor function), the parietal lobe (tactile sense), the temporal lobe (auditory perception), and the occipital lobe (visual sense).

The prefrontal region, which makes up a large part of the frontal lobe, is the distinctive feature of the human brain and is the foundation of creativity, memory, communication, and self-control.

The brain as seen from the left side

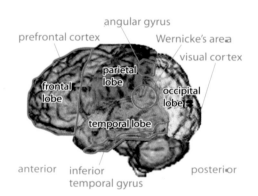

• **Note 3 :**

At the onset of this training your brain function will improve somewhat steadily. However, you will probably hit a wall at a certain point. You may wonder why your results are not showing any sign of improvement. But if you hang in there and continue your training, a breakthrough will come, and you will see your scores suddenly jump. If you are facing doldrums, just remember that your brain is preparing for a leap.

Day 1

Before you start, please take the Pre-training Prefrontal Cortex Evaluation.

Date ☐ M ☐ D

Start Time ☐ : ☐

$3 \times 5 =$ ☐

$4 + 5 =$ ☐

$10 - 6 =$ ☐

$3 + 9 =$ ☐

$3 \div 1 =$ ☐

$2 + 5 =$ ☐

$2 \times 3 =$ ☐

$5 \times 5 =$ ☐

$2 - 0 =$ ☐

$10 \div 7 =$ ☐ REMAINDER ☐

$3 \times 8 =$ ☐

$5 + 6 =$ ☐

$4 + 3 =$ ☐

$10 - 8 =$ ☐

$5 \div 1 =$ ☐

$1 \times 3 =$ ☐

$9 - 7 =$ ☐

$10 \div 5 =$ ☐

$2 + 7 =$ ☐

$14 - 5 =$ ☐

$3 \times 2 =$ ☐

$15 - 7 =$ ☐

$8 \times 5 =$ ☐

$3 \div 2 =$ ☐ REMAINDER ☐

$9 + 6 =$ ☐

$3 - 1 =$ ☐

$9 + 2 =$ ☐

$6 \times 8 =$ ☐

$11 - 5 =$ ☐

$7 \times 7 =$ ☐

$6 \div 3 =$ ☐

$10 \div 8 =$ ☐ REMAINDER ☐

$8 - 0 =$ ☐

$7 - 2 =$ ☐

$8 \times 0 =$ ☐

$4 - 3 =$ ☐

$7 \div 3 =$ ☐ REMAINDER ☐

$6 - 1 =$ ☐

$1 + 5 =$ ☐

$2 \div 1 =$ ☐

$7 + 9 =$ ☐

$11 - 6 =$ ☐

$3 + 4 =$ ☐

$5 \div 4 =$ ☐ REMAINDER ☐

$1 \times 2 =$ ☐

$9 \div 4 =$ ☐ REMAINDER ☐

$4 \times 9 =$ ☐

$16 \div 8 =$ ☐

$8 + 8 =$ ☐

$3 + 2 =$ ☐

$1 + 3 =$ ☐

$9 + 9 =$ ☐

$8 \div 5 =$ ☐ REMAINDER ☐

$7 + 0 =$ ☐

$11 - 3 =$ ☐

$4 \times 4 =$ ☐

$6 - 5 =$ ☐

$9 \times 6 =$ ☐

$18 \div 9 =$ ☐

$3 \times 1 =$ ☐

$11 - 2 =$ ☐

$12 \div 4 =$ ☐

$7 \times 9 =$ ☐

$3 \times 3 =$ ☐

$10 - 7 =$ ☐

$17 \div 8 =$ ☐ REMAINDER ☐

$16 - 8 =$ ☐

$6 + 7 =$ ☐

$8 - 4 =$ ☐

$13 \div 3 =$ ☐ REMAINDER ☐

$9 - 2 =$ ☐

$15 \div 3 =$ ☐

$0 + 6 =$ ☐

$7 \times 8 =$ ☐

$9 \times 3 =$ ☐

$6 + 6 =$ ☐

$8 \div 2 =$ ☐

$9 \times 7 =$ ☐

$15 \div 4 =$ ☐ REMAINDER ☐

$2 \times 8 =$ ☐

$5 - 3 =$ ☐

$9 - 5 =$ ☐

$16 \div 3 =$ ☐ REMAINDER ☐

$9 \times 8 =$ ☐

$14 \div 7 =$ ☐

$15 - 9 =$ ☐

$6 + 2 =$ ☐

$12 \div 3 =$ ☐

$9 \times 2 =$ ☐

$7 + 1 =$ ☐

$2 + 8 =$ ☐

$16 - 7 =$ ☐

$6 + 4 =$ ☐

$8 \times 2 =$ ☐

$6 + 8 =$ ☐

$13 \div 6 =$ ☐ REMAINDER ☐

$9 - 8 =$ ☐

$6 + 3 =$ ☐

$18 \div 6 =$ ☐

$0 \times 9 =$ ☐

End Time ☐ : ☐

Start Time ☐ : ☐

$1 + 6 =$ ☐

$16 \div 6 =$ ☐ REMAINDER ☐

$15 - 6 =$ ☐

$9 \div 6 =$ ☐ REMAINDER ☐

$9 \times 0 =$ ☐

$10 \div 2 =$ ☐

$5 + 5 =$ ☐

$12 - 3 =$ ☐

$7 + 4 =$ ☐

$10 - 9 =$ ☐

$14 - 8 =$ ☐

$9 \times 4 =$ ☐

$5 + 4 =$ ☐

$4 \div 2 =$ ☐

$13 - 6 =$ ☐

$7 - 4 =$ ☐

$2 + 1 =$ ☐

$1 + 9 =$ ☐

$6 \times 5 =$ ☐

$7 + 7 =$ ☐

$7 \div 1 =$ ☐

$6 \div 6 =$ ☐

$8 \div 6 =$ ☐ REMAINDER ☐

$7 \times 3 =$ ☐

$9 - 4 =$ ☐

$4 \times 3 =$ ☐

$6 \times 0 =$ ☐

$4 \times 6 =$ ☐

$17 \div 4 =$ ☐ REMAINDER ☐

$5 - 2 =$ ☐

$11 \div 5 =$ ☐ REMAINDER ☐

$7 - 5 =$ ☐

$4 \times 7 =$ ☐

$8 \times 3 =$ ☐

$3 + 6 =$ ☐

$6 \times 6 =$ ☐

$9 \times 1 =$ ☐

$3 - 0 =$ ☐

$8 + 5 =$ ☐

$8 - 2 =$ ☐

$4 + 1 =$ ☐

$12 \div 8 =$ ☐ REMAINDER ☐

$12 \div 6 =$ ☐

$5 + 9 =$ ☐

$1 + 2 =$ ☐

$4 + 2 =$ ☐

$2 \times 9 =$ ☐

$13 - 4 =$ ☐

16

$9 - 3 =$ ☐

$6 \div 2 =$ ☐

$16 \div 2 = $ ☐

$7 \times 2 = $ ☐

$2 + 4 = $ ☐

$9 \times 5 = $ ☐

$12 - 7 = $ ☐

$14 \div 2 = $ ☐

$10 - 3 = $ ☐

$8 + 6 = $ ☐

$5 \times 7 = $ ☐

$3 - 2 = $ ☐

$6 + 1 = $ ☐

$13 \div 8 = $ ☐ REMAINDER ☐

$1 \times 0 = $ ☐

$2 + 6 = $ ☐

$14 - 9 = $ ☐

$5 + 8 = $ ☐

$17 \div 2 = $ ☐ REMAINDER ☐

$8 + 2 = $ ☐

$5 \times 4 = $ ☐

$15 - 8 = $ ☐

$2 \times 6 = $ ☐

$6 - 0 = $ ☐

$14 \div 5 = $ ☐ REMAINDER ☐

$9 - 8 = $ ☐

$15 \div 5 = $ ☐

$4 + 9 = $ ☐

$11 - 7 = $ ☐

$8 \div 8 = $ ☐

$7 \times 5 = $ ☐

$3 \times 9 = $ ☐

$14 \div 8 = $ ☐ REMAINDER ☐

$9 - 5 = $ ☐

$7 \div 6 = $ ☐ REMAINDER ☐

$3 \times 7 = $ ☐

$8 \times 6 = $ ☐

$2 + 0 = $ ☐

$12 \div 2 = $ ☐

$8 - 6 = $ ☐

$10 - 4 = $ ☐

$1 + 1 = $ ☐

$8 + 0 = $ ☐

$4 \times 2 = $ ☐

$3 \times 6 = $ ☐

$18 \div 3 = $ ☐

$9 + 4 = $ ☐

$7 + 3 = $ ☐

$3 \times 4 = $ ☐

$5 \div 5 = $ ☐

$8 \div 3 = $ ☐ REMAINDER ☐

$7 - 1 = $ ☐

End Time ☐ : ☐

Day 3

Start Time ⬜ : ⬜

$15 \div 4 =$ ⬜ REMAINDER ⬜ $0 + 7 =$ ⬜

$8 - 7 =$ ⬜ $8 - 6 =$ ⬜ $3 + 4 =$ ⬜

$3 + 7 =$ ⬜ $0 \times 8 =$ ⬜ $9 \div 9 =$ ⬜

$12 \div 4 =$ ⬜ $8 + 1 =$ ⬜ $3 \times 8 =$ ⬜

$11 \div 5 =$ ⬜ REMAINDER ⬜ $8 \div 3 =$ ⬜ REMAINDER ⬜ $6 + 6 =$ ⬜

$14 - 5 =$ ⬜ $13 - 9 =$ ⬜ $9 \times 6 =$ ⬜

$2 + 9 =$ ⬜ $5 - 2 =$ ⬜ $4 \times 9 =$ ⬜

$3 \times 9 =$ ⬜ $3 \times 0 =$ ⬜ $12 \div 6 =$ ⬜

$10 \div 4 =$ ⬜ REMAINDER ⬜ $10 - 2 =$ ⬜ $8 - 3 =$ ⬜

$9 + 6 =$ ⬜ $9 \times 4 =$ ⬜ $3 \times 6 =$ ⬜

$5 - 0 =$ ⬜ $7 \div 1 =$ ⬜ $13 - 6 =$ ⬜

$8 \times 1 =$ ⬜ $2 + 6 =$ ⬜ $12 \div 2 =$ ⬜

$9 \div 5 =$ ⬜ REMAINDER ⬜ $5 \div 3 =$ ⬜ REMAINDER ⬜ $13 - 7 =$ ⬜

$2 \times 6 =$ ⬜ $7 + 1 =$ ⬜ $6 \times 9 =$ ⬜

$1 + 7 =$ ⬜ $1 + 4 =$ ⬜ $3 - 3 =$ ⬜

$15 - 9 =$ ⬜ $2 - 2 =$ ⬜ $6 + 7 =$ ⬜

$8 \times 0 =$ ⬜ $8 + 5 =$ ⬜ $4 \div 2 =$ ⬜

$11 - 9 =$ ☐ $8 - 4 =$ ☐ $2 + 8 =$ ☐

$7 - 0 =$ ☐ $0 + 8 =$ ☐ $7 + 6 =$ ☐

$3 + 6 =$ ☐ $2 \times 9 =$ ☐ $6 - 4 =$ ☐

$4 \times 5 =$ ☐ $5 \div 5 =$ ☐ $7 \times 8 =$ ☐

$9 \times 8 =$ ☐ $3 \times 5 =$ ☐ $18 \div 9 =$ ☐

$6 \div 2 =$ ☐ $1 + 3 =$ ☐ $10 - 7 =$ ☐

$1 \times 5 =$ ☐ $12 - 5 =$ ☐ $11 \div 8 =$ ☐ REMAINDER ☐

$9 - 8 =$ ☐ $7 \times 2 =$ ☐ $7 + 4 =$ ☐

$12 - 3 =$ ☐ $6 \div 3 =$ ☐ $9 \times 9 =$ ☐

$5 + 1 =$ ☐ $16 \div 7 =$ ☐ REMAINDER ☐ $12 \div 8 =$ ☐ REMAINDER ☐

$5 \times 0 =$ ☐ $6 + 4 =$ ☐ $5 \times 4 =$ ☐

$9 + 8 =$ ☐ $4 + 7 =$ ☐ $15 - 8 =$ ☐

$17 \div 7 =$ ☐ REMAINDER ☐ $18 \div 3 =$ ☐ $9 \times 2 =$ ☐

$5 + 4 =$ ☐ $5 \times 8 =$ ☐ $7 \div 4 =$ ☐ REMAINDER ☐

$8 \div 6 =$ ☐ REMAINDER ☐ $10 - 4 =$ ☐ $10 \div 5 =$ ☐

$2 \times 8 =$ ☐ $8 - 0 =$ ☐ $7 - 1 =$ ☐

$9 \div 3 =$ ☐ $1 + 5 =$ ☐   End Time ☐ : ☐

Start Time ☐ : ☐

$5 \div 2 =$ ☐ REMAINDER ☐

$10 - 1 =$ ☐

$1 \times 3 =$ ☐

$3 - 1 =$ ☐

$4 + 8 =$ ☐

$9 \times 5 =$ ☐

$6 - 1 =$ ☐

$9 + 4 =$ ☐

$3 + 3 =$ ☐

$6 \div 6 =$ ☐

$6 \times 2 =$ ☐

$2 + 0 =$ ☐

$2 \times 1 =$ ☐

$9 + 1 =$ ☐

$15 \div 5 =$ ☐

$9 - 2 =$ ☐

$8 \times 5 =$ ☐

$4 - 3 =$ ☐

$8 \times 4 =$ ☐

$4 + 2 =$ ☐

$7 \div 5 =$ ☐ REMAINDER ☐

$18 \div 2 =$ ☐

$8 + 3 =$ ☐

$1 \times 0 =$ ☐

$12 \div 5 =$ ☐ REMAINDER ☐

$6 - 2 =$ ☐

$16 - 8 =$ ☐

$6 + 2 =$ ☐

$14 - 7 =$ ☐

$2 \times 4 =$ ☐

$6 \div 5 =$ ☐ REMAINDER ☐

$1 - 1 =$ ☐

$13 \div 6 =$ ☐ REMAINDER ☐

$7 \times 3 =$ ☐

$6 + 1 =$ ☐

$11 - 5 =$ ☐

$8 \times 3 =$ ☐

$13 \div 8 =$ ☐ REMAINDER ☐

$4 \times 3 =$ ☐

$3 - 2 =$ ☐

$7 \div 7 =$ ☐

$4 \times 8 =$ ☐

$11 - 3 =$ ☐

$2 + 7 =$ ☐

$14 \div 2 =$ ☐

$1 + 9 =$ ☐

$17 - 9 =$ ☐

$3 + 4 =$ ☐

$16 \div 4 =$ ☐

$8 + 9 =$ ☐

$8 \times 6 =$ ☐

$4 \div 4 =$ ☐

$12 - 4 =$ ☐

$3 \times 2 =$ ☐

$0 + 6 =$ ☐

$7 - 0 =$ ☐

$8 + 2 =$ ☐

$14 \div 9 =$ ☐ REMAINDER ☐

$6 - 5 =$ ☐

$9 + 9 =$ ☐

$8 \div 7 =$ ☐ REMAINDER ☐

$18 \div 6 =$ ☐

$5 \times 2 =$ ☐

$8 + 6 =$ ☐

$10 \div 6 =$ ☐ REMAINDER ☐

$1 + 1 =$ ☐

$3 \times 3 =$ ☐

$8 \div 1 =$ ☐

$15 - 7 =$ ☐

$6 + 3 =$ ☐

$6 \times 6 =$ ☐

$11 - 4 =$ ☐

$15 \div 8 =$ ☐ REMAINDER ☐

$13 - 4 =$ ☐

$2 \times 3 =$ ☐

$16 - 7 =$ ☐

$8 \div 4 =$ ☐

$9 \times 3 =$ ☐

$5 \div 1 =$ ☐

$8 + 7 =$ ☐

$7 \times 9 =$ ☐

$2 + 2 =$ ☐

$7 - 5 =$ ☐

$1 \times 7 =$ ☐

$9 \div 2 =$ ☐ REMAINDER ☐

$3 \times 6 =$ ☐

$8 - 4 =$ ☐

$4 + 3 =$ ☐

$1 + 8 =$ ☐

$18 - 9 =$ ☐

$8 - 8 =$ ☐

$11 \div 9 =$ ☐ REMAINDER ☐

$2 \div 1 =$ ☐

$4 \times 6 =$ ☐

$6 + 8 =$ ☐

$1 \times 2 =$ ☐

$15 \div 3 =$ ☐

$8 - 5 =$ ☐

$5 \times 9 =$ ☐

$7 + 8 =$ ☐

End Time ☐ : ☐

Start Time ☐ : ☐

7 − 6 = ☐

12 ÷ 2 = ☐

2 − 1 = ☐

8 × 1 = ☐

0 + 6 = ☐

5 + 3 = ☐

15 ÷ 3 = ☐

4 + 7 = ☐

19 ÷ 6 = ☐ REMAINDER ☐

3 × 4 = ☐

9 × 9 = ☐

1 + 7 = ☐

6 − 3 = ☐

12 ÷ 8 = ☐ REMAINDER ☐

5 × 5 = ☐

9 − 6 = ☐

9 + 2 = ☐

7 × 0 = ☐

3 ÷ 2 = ☐ REMAINDER ☐

12 − 9 = ☐

16 − 9 = ☐

1 + 6 = ☐

3 × 6 = ☐

8 − 8 = ☐

6 ÷ 3 = ☐

10 ÷ 8 = ☐ REMAINDER ☐

13 − 7 = ☐

6 + 6 = ☐

13 − 6 = ☐

7 × 9 = ☐

16 ÷ 8 = ☐

7 + 8 = ☐

3 × 3 = ☐

4 × 7 = ☐

5 × 8 = ☐

2 + 9 = ☐

2 ÷ 2 = ☐

2 + 4 = ☐

3 − 2 = ☐

8 + 5 = ☐

6 × 7 = ☐

17 ÷ 2 = ☐ REMAINDER ☐

11 − 6 = ☐

7 ÷ 5 = ☐ REMAINDER ☐

6 − 5 = ☐

4 + 1 = ☐

10 ÷ 5 = ☐

10 − 1 = ☐

4 + 2 = ☐

9 × 6 = ☐

$5 \times 6 =$ ☐

$3 \times 2 =$ ☐

$8 - 2 =$ ☐

$2 \div 1 =$ ☐

$17 \div 6 =$ ☐ REMAINDER ☐

$3 + 9 =$ ☐

$17 - 9 =$ ☐

$2 + 6 =$ ☐

$6 \times 9 =$ ☐

$7 \div 3 =$ ☐ REMAINDER ☐

$11 - 9 =$ ☐

$5 - 0 =$ ☐

$7 + 7 =$ ☐

$2 \times 6 =$ ☐

$4 \div 2 =$ ☐

$2 \times 4 =$ ☐

$7 + 5 =$ ☐

$5 \div 2 =$ ☐ REMAINDER ☐

$8 \times 2 =$ ☐

$7 + 6 =$ ☐

$12 - 7 =$ ☐

$7 - 2 =$ ☐

$3 \times 8 =$ ☐

$1 + 4 =$ ☐

$18 \div 7 =$ ☐ REMAINDER ☐

$15 \div 5 =$ ☐

$9 - 5 =$ ☐

$11 \div 7 =$ ☐ REMAINDER ☐

$14 - 8 =$ ☐

$12 \div 3 =$ ☐

$5 + 4 =$ ☐

$7 \times 6 =$ ☐

$3 + 7 =$ ☐

$8 - 3 =$ ☐

$6 + 2 =$ ☐

$9 + 4 =$ ☐

$11 - 3 =$ ☐

$4 \times 9 =$ ☐

$6 + 3 =$ ☐

$4 \times 5 =$ ☐

$16 \div 4 =$ ☐

$6 \times 8 =$ ☐

$1 + 5 =$ ☐

$4 - 4 =$ ☐

$10 - 2 =$ ☐

$18 \div 6 =$ ☐

$1 \times 4 =$ ☐

$7 \times 8 =$ ☐

$8 \div 3 =$ ☐ REMAINDER ☐

$18 \div 9 =$ ☐

End Time ☐ : ☐

22

I. Counting Test

Measure the time required for you to count from 1 to 120 aloud as fast as you can.

☐ sec.

II. Word Memorization Test

Memorize as many words as you can **within two minutes.**

friend	rumour	clock	snore	storm	pair
orphan	criticism	seat	piggyback	period	brick
back	fledgling	delay	collapse	ladder	health
Spanish	sword	cupboard	raindrop	vegetable	duckling
sample	plant	party	spiral	clothing	revision

Write out as many words as you can remember **in two minutes** on the back of this page. How many words can you remember?

Number of words memorized ☐ words

Word Memorization Test Answers

III. Stroop Test

Please take the **Stroop Test** for Week 1, located on page **iv** of the Appendix.

Start Time ▢ : ▢

$14 - 7 =$ ▢

$8 \times 9 =$ ▢

$11 \div 9 =$ ▢ REMAINDER ▢

$9 \times 7 =$ ▢

$3 + 6 =$ ▢

$12 - 5 =$ ▢

$7 \div 1 =$ ▢

$3 \times 9 =$ ▢

$6 + 7 =$ ▢

$7 - 1 =$ ▢

$6 - 2 =$ ▢

$8 \times 1 =$ ▢

$2 + 7 =$ ▢

$16 \div 5 =$ ▢ REMAINDER ▢

$4 - 3 =$ ▢

$12 \div 4 =$ ▢

$15 \div 9 =$ ▢ REMAINDER ▢

$4 \times 8 =$ ▢

$8 + 3 =$ ▢

$13 - 4 =$ ▢

$18 \div 2 =$ ▢

$8 + 1 =$ ▢

$14 - 6 =$ ▢

$12 \div 7 =$ ▢ REMAINDER ▢

$2 \times 9 =$ ▢

$9 + 3 =$ ▢

$4 + 8 =$ ▢

$0 \times 2 =$ ▢

$10 - 6 =$ ▢

$5 + 2 =$ ▢

$8 - 1 =$ ▢

$13 \div 4 =$ ▢ REMAINDER ▢

$7 \times 7 =$ ▢

$14 \div 7 =$ ▢

$9 + 7 =$ ▢

$18 - 9 =$ ▢

$18 \div 3 =$ ▢

$3 \times 7 =$ ▢

$9 + 8 =$ ▢

$5 \times 2 =$ ▢

$7 + 2 =$ ▢

$17 \div 7 =$ ▢ REMAINDER ▢

$3 + 1 =$ ▢

$1 - 1 =$ ▢

$6 - 4 =$ ▢

$8 \div 2 =$ ▢

$9 + 0 =$ ▢

$2 \times 5 =$ ▢

$4 \times 2 =$ ▢

$9 - 7 =$ ▢

$3 + 2 =$ ☐

$7 - 3 =$ ☐

$9 \div 3 =$ ☐

$5 \times 1 =$ ☐

$10 - 4 =$ ☐

$2 + 2 =$ ☐

$5 - 3 =$ ☐

$18 \div 5 =$ ☐ REMAINDER ☐

$6 \times 5 =$ ☐

$7 \times 3 =$ ☐

$5 - 0 =$ ☐

$9 \times 5 =$ ☐

$8 + 2 =$ ☐

$8 - 3 =$ ☐

$6 \div 2 =$ ☐

$3 + 8 =$ ☐

$10 \div 2 =$ ☐

$6 \div 6 =$ ☐

$6 + 8 =$ ☐

$3 \times 1 =$ ☐

$7 \div 6 =$ ☐ REMAINDER ☐

$11 - 5 =$ ☐

$1 + 3 =$ ☐

$2 \times 3 =$ ☐

$5 \div 3 =$ ☐ REMAINDER ☐

$8 \div 5 =$ ☐ REMAINDER ☐

$7 + 3 =$ ☐

$4 \times 6 =$ ☐

$0 + 3 =$ ☐

$19 \div 4 =$ ☐ REMAINDER ☐

$14 - 5 =$ ☐

$4 \times 4 =$ ☐

$11 - 4 =$ ☐

$3 \times 5 =$ ☐

$16 - 8 =$ ☐

$7 \times 4 =$ ☐

$9 - 1 =$ ☐

$4 + 5 =$ ☐

$14 \div 2 =$ ☐

$8 + 7 =$ ☐

$7 \times 5 =$ ☐

$15 - 7 =$ ☐

$12 \div 6 =$ ☐

$6 \times 4 =$ ☐

$7 + 1 =$ ☐

$16 \div 2 =$ ☐

$7 + 9 =$ ☐

$19 \div 7 =$ ☐ REMAINDER ☐

$2 \times 2 =$ ☐

$7 - 5 =$ ☐

End Time ☐ : ☐

Start Time [ ] : [ ]

7 + 9 =

13 − 5 =

4 × 7 =

19 ÷ 2 = [ ] REMAINDER [ ]

5 + 1 =

8 − 7 =

17 − 9 =

3 ÷ 2 = [ ] REMAINDER [ ]

2 + 8 =

14 ÷ 7 =

4 + 9 =

4 × 8 =

9 + 7 =

0 × 1 =

18 ÷ 6 =

6 × 6 =

0 × 4 =

12 ÷ 3 =

5 + 3 =

7 × 5 =

2 + 6 =

7 − 2 =

11 − 3 =

8 + 4 =

18 ÷ 8 = [ ] REMAINDER [ ]

6 − 1 =

1 × 6 =

13 ÷ 3 = [ ] REMAINDER [ ]

9 − 2 =

5 + 2 =

2 × 9 =

6 − 5 =

14 ÷ 9 = [ ] REMAINDER [ ]

9 ÷ 3 =

5 × 6 =

15 − 8 =

2 + 7 =

17 − 8 =

1 + 2 =

7 × 9 =

6 − 3 =

6 + 7 =

8 ÷ 2 =

4 − 1 =

7 ÷ 1 =

13 − 7 =

9 × 7 =

3 + 4 =

6 × 8 =

5 ÷ 2 = [ ] REMAINDER [ ]

$2 + 4 =$ [ ]

$5 - 3 =$ [ ]

$14 \div 2 =$ [ ]

$3 \times 6 =$ [ ]

$13 - 9 =$ [ ]

$2 + 5 =$ [ ]

$13 \div 7 =$ [ ] REMAINDER [ ]

$3 + 6 =$ [ ]

$13 - 8 =$ [ ]

$10 \div 4 =$ [ ] REMAINDER [ ]

$8 + 1 =$ [ ]

$7 \times 3 =$ [ ]

$11 \div 9 =$ [ ] REMAINDER [ ]

$5 \times 9 =$ [ ]

$8 + 5 =$ [ ]

$5 \times 4 =$ [ ]

$8 - 3 =$ [ ]

$12 \div 6 =$ [ ]

$2 \times 3 =$ [ ]

$8 - 5 =$ [ ]

$8 \times 9 =$ [ ]

$5 \times 7 =$ [ ]

$10 \div 2 =$ [ ]

$8 + 2 =$ [ ]

$3 - 1 =$ [ ]

$18 \div 4 =$ [ ] REMAINDER [ ]

$11 - 6 =$ [ ]

$7 + 5 =$ [ ]

$5 \div 1 =$ [ ]

$6 + 1 =$ [ ]

$9 \times 9 =$ [ ]

$8 \times 3 =$ [ ]

$18 \div 3 =$ [ ]

$10 - 3 =$ [ ]

$4 \times 2 =$ [ ]

$16 \div 4 =$ [ ]

$7 + 8 =$ [ ]

$4 - 2 =$ [ ]

$3 + 8 =$ [ ]

$11 - 8 =$ [ ]

$18 - 9 =$ [ ]

$17 \div 9 =$ [ ] REMAINDER [ ]

$6 + 8 =$ [ ]

$15 \div 4 =$ [ ] REMAINDER [ ]

$7 \times 1 =$ [ ]

$10 \div 5 =$ [ ]

$0 \times 2 =$ [ ]

$7 - 3 =$ [ ]

$8 \times 2 =$ [ ]

$3 + 5 =$ [ ]

End Time [ ] : [ ]

Start Time ☐ : ☐

$5 \times 2 =$ ☐

$7 + 6 =$ ☐

$12 \div 2 =$ ☐

$3 \times 8 =$ ☐

$14 \div 6 =$ ☐ REMAINDER ☐

$6 + 4 =$ ☐

$12 - 8 =$ ☐

$1 + 8 =$ ☐

$2 \times 6 =$ ☐

$4 \div 2 =$ ☐

$18 \div 2 =$ ☐

$5 - 4 =$ ☐

$8 + 9 =$ ☐

$3 \times 3 =$ ☐

$9 + 5 =$ ☐

$7 - 6 =$ ☐

$12 - 6 =$ ☐

$4 \div 3 =$ ☐ REMAINDER ☐

$19 \div 8 =$ ☐ REMAINDER ☐

$0 + 9 =$ ☐

$8 + 7 =$ ☐

$4 + 2 =$ ☐

$8 - 1 =$ ☐

$9 \times 3 =$ ☐

$9 \div 7 =$ ☐ REMAINDER ☐

$6 \div 2 =$ ☐

$1 - 1 =$ ☐

$1 + 6 =$ ☐

$8 \times 4 =$ ☐

$5 + 7 =$ ☐

$6 \times 9 =$ ☐

$17 \div 5 =$ ☐ REMAINDER ☐

$6 \times 3 =$ ☐

$12 - 4 =$ ☐

$3 + 1 =$ ☐

$1 \times 4 =$ ☐

$9 \times 0 =$ ☐

$9 - 8 =$ ☐

$8 - 6 =$ ☐

$7 + 2 =$ ☐

$2 \times 5 =$ ☐

$16 \div 2 =$ ☐

$9 - 6 =$ ☐

$12 - 9 =$ ☐

$10 - 2 =$ ☐

$15 \div 3 =$ ☐

$11 \div 5 =$ ☐ REMAINDER ☐

$9 + 3 =$ ☐

$12 - 5 =$ ☐

$9 \times 6 =$ ☐

Time Required ☐ : ☐

6 + 6 = ☐   8 − 5 = ☐   16 − 9 = ☐

16 − 7 = ☐   4 × 5 = ☐   2 × 2 = ☐

9 − 4 = ☐   16 ÷ 5 = ☐ REMAINDER ☐   8 − 0 = ☐

12 ÷ 6 = ☐   12 − 7 = ☐   7 + 3 = ☐

5 × 3 = ☐   9 + 9 = ☐   6 ÷ 6 = ☐

7 × 4 = ☐   3 × 9 = ☐   9 × 4 = ☐

11 − 7 = ☐   12 ÷ 4 = ☐   11 ÷ 3 = ☐ REMAINDER ☐

7 ÷ 2 = ☐ REMAINDER ☐   1 + 7 = ☐   4 + 0 = ☐

9 + 0 = ☐   8 ÷ 4 = ☐   6 ÷ 3 = ☐

16 ÷ 8 = ☐   9 + 8 = ☐   6 + 9 = ☐

7 + 1 = ☐   13 ÷ 6 = ☐ REMAINDER ☐   7 × 6 = ☐

4 × 4 = ☐   3 × 7 = ☐   2 + 3 = ☐

2 − 1 = ☐   4 + 5 = ☐   17 ÷ 4 = ☐ REMAINDER ☐

3 ÷ 3 = ☐   14 − 6 = ☐   8 × 8 = ☐

2 × 8 = ☐   18 ÷ 7 = ☐ REMAINDER ☐   7 − 3 = ☐

7 − 4 = ☐   10 − 4 = ☐   2 + 9 = ☐

3 × 2 = ☐   3 × 5 = ☐

End Time ☐ : ☐

Start Time ☐ : ☐

$7 \div 3 =$ ☐ REMAINDER ☐

$4 \times 7 =$ ☐

$10 - 6 =$ ☐

$4 + 2 =$ ☐

$8 \div 4 =$ ☐

$6 \times 4 =$ ☐

$5 + 3 =$ ☐

$8 - 1 =$ ☐

$9 + 6 =$ ☐

$5 \div 5 =$ ☐

$3 \times 7 =$ ☐

$16 \div 8 =$ ☐

$6 - 3 =$ ☐

$1 + 2 =$ ☐

$12 - 7 =$ ☐

$2 \times 2 =$ ☐

$14 \div 7 =$ ☐

$9 + 2 =$ ☐

$1 - 1 =$ ☐

$1 + 3 =$ ☐

$2 \times 9 =$ ☐

$5 \div 1 =$ ☐

$10 - 9 =$ ☐

$17 - 8 =$ ☐

$8 \times 8 =$ ☐

$3 + 9 =$ ☐

$9 \times 1 =$ ☐

$7 \times 5 =$ ☐

$3 + 6 =$ ☐

$9 + 8 =$ ☐

$14 \div 6 =$ ☐ REMAINDER ☐

$12 - 8 =$ ☐

$15 \div 4 =$ ☐ REMAINDER ☐

$9 \times 9 =$ ☐

$15 \div 7 =$ ☐ REMAINDER ☐

$2 + 9 =$ ☐

$5 - 1 =$ ☐

$9 \times 3 =$ ☐

$7 + 3 =$ ☐

$8 - 2 =$ ☐

$10 \div 6 =$ ☐ REMAINDER ☐

$3 \times 5 =$ ☐

$17 \div 8 =$ ☐ REMAINDER ☐

$4 - 1 =$ ☐

$2 + 6 =$ ☐

$3 \times 8 =$ ☐

$15 \div 5 =$ ☐

$16 - 7 =$ ☐

$4 + 1 =$ ☐

$4 - 3 =$ ☐

$3 + 5 =$ ☐

$19 \div 4 =$ ☐ REMAINDER ☐

$4 \times 1 =$ ☐

$12 \div 5 =$ ☐ REMAINDER ☐

$6 \times 9 =$ ☐

$10 - 4 =$ ☐

$18 \div 9 =$ ☐

$3 - 1 =$ ☐

$4 + 5 =$ ☐

$2 \times 3 =$ ☐

$9 \div 9 =$ ☐

$9 - 3 =$ ☐

$7 \times 8 =$ ☐

$8 + 1 =$ ☐

$3 \div 3 =$ ☐

$6 + 2 =$ ☐

$10 - 2 =$ ☐

$5 \times 2 =$ ☐

$4 + 9 =$ ☐

$1 + 5 =$ ☐

$18 - 9 =$ ☐

$4 \times 9 =$ ☐

$7 \div 5 =$ ☐ REMAINDER ☐

$4 + 8 =$ ☐

$11 - 6 =$ ☐

$12 \div 3 =$ ☐

$7 \times 9 =$ ☐

$3 - 2 =$ ☐

$5 + 7 =$ ☐

$4 - 4 =$ ☐

$9 \times 7 =$ ☐

$19 \div 7 =$ ☐ REMAINDER ☐

$6 - 5 =$ ☐

$7 \div 7 =$ ☐

$8 + 5 =$ ☐

$3 \times 1 =$ ☐

$6 \div 5 =$ ☐ REMAINDER ☐

$5 - 3 =$ ☐

$15 \div 3 =$ ☐

$4 \times 2 =$ ☐

$7 + 1 =$ ☐

$3 \times 9 =$ ☐

$8 + 8 =$ ☐

$3 \div 1 =$ ☐

$10 - 5 =$ ☐

$6 + 9 =$ ☐

$1 \times 2 =$ ☐

$11 - 7 =$ ☐

$4 \times 5 =$ ☐

$9 \div 7 =$ ☐ REMAINDER ☐

End Time ☐ : ☐

Start Time ☐ : ☐

10 − 3 = ☐

3 + 3 = ☐

17 − 9 = ☐

5 ÷ 3 = ☐ REMAINDER ☐

8 − 5 = ☐

8 + 7 = ☐

6 × 3 = ☐

9 + 5 = ☐

6 − 2 = ☐

7 × 6 = ☐

5 + 5 = ☐

9 ÷ 2 = ☐ REMAINDER ☐

9 − 7 = ☐

6 + 3 = ☐

14 ÷ 2 = ☐

8 × 5 = ☐

6 + 4 = ☐

5 − 4 = ☐

18 ÷ 2 = ☐

1 + 6 = ☐

4 + 2 = ☐

12 − 9 = ☐

5 × 7 = ☐

16 ÷ 4 = ☐

7 × 2 = ☐

9 − 5 = ☐

13 ÷ 6 = ☐ REMAINDER ☐

5 × 3 = ☐

10 ÷ 2 = ☐

5 − 2 = ☐

7 + 8 = ☐

3 × 4 = ☐

6 ÷ 4 = ☐ REMAINDER ☐

4 + 3 = ☐

10 ÷ 7 = ☐ REMAINDER ☐

7 − 1 = ☐

8 × 9 = ☐

2 × 6 = ☐

10 ÷ 4 = ☐ REMAINDER ☐

12 − 6 = ☐

11 − 4 = ☐

2 + 7 = ☐

16 ÷ 2 = ☐

8 × 6 = ☐

9 × 4 = ☐

18 ÷ 3 = ☐

12 − 4 = ☐

9 + 7 = ☐

3 + 2 = ☐

5 × 9 = ☐

$18 \div 7 =$ ☐ REMAINDER ☐ $8 \times 1 =$ ☐ $7 + 6 =$ ☐

$0 \times 9 =$ ☐ $7 - 5 =$ ☐ $6 \times 6 =$ ☐

$7 \times 3 =$ ☐ $19 \div 6 =$ ☐ REMAINDER ☐ $10 \div 5 =$ ☐

$4 - 4 =$ ☐ $2 + 3 =$ ☐ $1 + 4 =$ ☐

$10 - 7 =$ ☐ $7 + 4 =$ ☐ $1 \times 3 =$ ☐

$7 + 7 =$ ☐ $9 - 6 =$ ☐ $8 \div 8 =$ ☐

$16 - 8 =$ ☐ $11 \div 8 =$ ☐ REMAINDER ☐ $14 - 7 =$ ☐

$2 \div 1 =$ ☐ $6 - 5 =$ ☐ $18 \div 6 =$ ☐

$11 - 9 =$ ☐ $8 + 3 =$ ☐ $14 - 8 =$ ☐

$9 + 1 =$ ☐ $4 - 2 =$ ☐ $5 + 2 =$ ☐

$17 \div 2 =$ ☐ REMAINDER ☐ $2 + 4 =$ ☐ $6 \times 8 =$ ☐

$1 + 7 =$ ☐ $2 \times 7 =$ ☐ $5 \times 5 =$ ☐

$18 \div 4 =$ ☐ REMAINDER ☐ $5 \times 4 =$ ☐ $8 - 6 =$ ☐

$6 \times 7 =$ ☐ $12 \div 4 =$ ☐ $9 \times 8 =$ ☐

$9 \div 3 =$ ☐ $8 \div 2 =$ ☐ $1 + 9 =$ ☐

$4 \times 8 =$ ☐ $11 - 3 =$ ☐ $11 \div 7 =$ ☐ REMAINDER ☐

$4 + 4 =$ ☐ $8 \times 3 =$ ☐   **End Time** ☐ : ☐

I. Counting Test

Measure the time required for you to count from 1 to 120 aloud as fast as you can.

☐ sec.

II. Word Memorization Test

Memorize as many words as you can **within two minutes**.

bud	countryside	toothpick	gesture	birdbox	auction
truth	partner	chick	son	theatre	life
stomach	paintbrush	sheep	scales	maple	shrine
stroll	weight	eyelashes	oyster	world	family
woman	keepsake	echo	kite	dolphin	peony

Write out as many words as you can remember **in two minutes** on the back of this page. How many words can you remember?

Number of words memorized ☐ words

Word Memorization Test Answers

III. Stroop Test

Please take the **Stroop Test** for Week 2, located on page **v** of the Appendix.

Start Time ☐ : ☐

$18 \div 3 =$ ☐

$7 \times 7 =$ ☐

$14 \div 7 =$ ☐

$1 + 4 =$ ☐

$6 \div 4 =$ ☐ REMAINDER ☐

$11 - 8 =$ ☐

$9 - 4 =$ ☐

$3 \times 8 =$ ☐

$5 + 4 =$ ☐

$6 + 3 =$ ☐

$9 \times 6 =$ ☐

$3 + 5 =$ ☐

$1 \times 2 =$ ☐

$14 \div 3 =$ ☐ REMAINDER ☐

$9 - 5 =$ ☐

$3 - 2 =$ ☐

$8 \div 7 =$ ☐ REMAINDER ☐

$5 \times 3 =$ ☐

$3 + 1 =$ ☐

$2 - 1 =$ ☐

$5 \div 5 =$ ☐

$6 \times 4 =$ ☐

$9 \times 7 =$ ☐

$4 + 4 =$ ☐

$1 - 1 =$ ☐

$17 \div 5 =$ ☐ REMAINDER ☐

$9 \times 9 =$ ☐

$18 \div 2 =$ ☐

$9 + 8 =$ ☐

$6 - 6 =$ ☐

$7 \times 1 =$ ☐

$10 - 7 =$ ☐

$2 + 9 =$ ☐

$10 - 3 =$ ☐

$6 \times 6 =$ ☐

$4 + 5 =$ ☐

$7 - 6 =$ ☐

$9 + 4 =$ ☐

$11 \div 6 =$ ☐ REMAINDER ☐

$3 + 8 =$ ☐

$8 \div 2 =$ ☐

$10 - 6 =$ ☐

$5 \times 5 =$ ☐

$7 + 6 =$ ☐

$11 - 9 =$ ☐

$14 \div 4 =$ ☐ REMAINDER ☐

$1 \times 9 =$ ☐

$6 \div 3 =$ ☐

$15 - 8 =$ ☐

$6 + 8 =$ ☐

$1 + 3 =$ ☐ $8 \div 4 =$ ☐ $4 \times 4 =$ ☐

$4 \times 9 =$ ☐ $5 + 9 =$ ☐ $8 - 4 =$ ☐

$4 \div 2 =$ ☐ $7 + 3 =$ ☐ $9 \div 1 =$ ☐

$14 - 6 =$ ☐ $6 \times 8 =$ ☐ $5 \times 7 =$ ☐

$7 \times 4 =$ ☐ $9 - 1 =$ ☐ $3 + 7 =$ ☐

$5 + 6 =$ ☐ $11 \div 7 =$ ☐ REMAINDER ☐ $5 + 0 =$ ☐

$12 - 8 =$ ☐ $1 \times 3 =$ ☐ $12 \div 4 =$ ☐

$6 \div 6 =$ ☐ $16 \div 6 =$ ☐ REMAINDER ☐ $8 - 3 =$ ☐

$18 \div 8 =$ ☐ REMAINDER ☐ $11 - 2 =$ ☐ $12 - 4 =$ ☐

$9 \times 3 =$ ☐ $1 \times 4 =$ ☐ $9 \times 2 =$ ☐

$9 + 9 =$ ☐ $4 + 2 =$ ☐ $19 \div 4 =$ ☐ REMAINDER ☐

$8 - 8 =$ ☐ $9 \times 8 =$ ☐ $9 - 6 =$ ☐

$12 \div 6 =$ ☐ $2 + 8 =$ ☐ $4 \times 3 =$ ☐

$4 \times 8 =$ ☐ $15 - 7 =$ ☐ $10 \div 9 =$ ☐ REMAINDER ☐

$5 - 1 =$ ☐ $18 - 9 =$ ☐ $4 + 1 =$ ☐

$14 \div 6 =$ ☐ REMAINDER ☐ $7 \div 7 =$ ☐ $1 + 7 =$ ☐

$1 + 2 =$ ☐ $2 \times 9 =$ ☐   **End Time** ☐ : ☐

38

Day 12

Start Time ☐ : ☐

$5 + 8 =$ ☐

$14 - 7 =$ ☐

$15 \div 8 =$ ☐ REMAINDER ☐

$6 + 2 =$ ☐

$7 \times 9 =$ ☐

$9 - 3 =$ ☐

$18 \div 9 =$ ☐

$5 + 1 =$ ☐

$2 \times 9 =$ ☐

$3 \times 1 =$ ☐

$9 - 2 =$ ☐

$3 + 6 =$ ☐

$11 \div 9 =$ ☐ REMAINDER ☐

$3 \div 1 =$ ☐

$7 \times 5 =$ ☐

$11 - 6 =$ ☐

$7 \times 2 =$ ☐

$1 + 5 =$ ☐

$2 \times 1 =$ ☐

$4 \div 4 =$ ☐

$6 - 5 =$ ☐

$17 - 9 =$ ☐

$6 + 9 =$ ☐

$13 \div 2 =$ ☐ REMAINDER ☐

$12 - 9 =$ ☐

$4 \div 1 =$ ☐

$2 + 3 =$ ☐

$5 \times 8 =$ ☐

$4 \times 7 =$ ☐

$16 \div 2 =$ ☐

$6 - 1 =$ ☐

$8 - 5 =$ ☐

$4 + 6 =$ ☐

$7 + 4 =$ ☐

$4 - 1 =$ ☐

$6 \times 3 =$ ☐

$7 \div 6 =$ ☐ REMAINDER ☐

$7 + 8 =$ ☐

$16 \div 8 =$ ☐

$8 + 9 =$ ☐

$3 \times 4 =$ ☐

$2 + 6 =$ ☐

$16 \div 7 =$ ☐ REMAINDER ☐

$4 - 2 =$ ☐

$0 + 5 =$ ☐

$9 \div 7 =$ ☐ REMAINDER ☐

$10 - 2 =$ ☐

$15 - 6 =$ ☐

$1 \times 7 =$ ☐

$7 \times 6 =$ ☐

$7 - 2 =$ [ ]

$5 \div 4 =$ [ ] REMAINDER [ ]

$16 - 7 =$ [ ]

$6 \div 2 =$ [ ]

$8 \times 2 =$ [ ]

$6 \times 9 =$ [ ]

$5 + 7 =$ [ ]

$2 + 5 =$ [ ]

$9 \div 3 =$ [ ]

$11 - 3 =$ [ ]

$4 \times 5 =$ [ ]

$4 + 9 =$ [ ]

$19 \div 6 =$ [ ] REMAINDER [ ]

$9 + 2 =$ [ ]

$5 - 2 =$ [ ]

$6 + 7 =$ [ ]

$6 \times 2 =$ [ ]

$7 + 7 =$ [ ]

$10 \div 7 =$ [ ] REMAINDER [ ]

$2 \times 7 =$ [ ]

$10 - 4 =$ [ ]

$2 \times 3 =$ [ ]

$8 - 4 =$ [ ]

$15 \div 5 =$ [ ]

$6 - 3 =$ [ ]

$5 - 1 =$ [ ]

$2 + 1 =$ [ ]

$8 \times 8 =$ [ ]

$2 \times 5 =$ [ ]

$14 \div 9 =$ [ ] REMAINDER [ ]

$2 \times 6 =$ [ ]

$7 + 5 =$ [ ]

$6 \div 1 =$ [ ]

$16 \div 4 =$ [ ]

$3 - 1 =$ [ ]

$8 \times 3 =$ [ ]

$10 - 8 =$ [ ]

$10 \div 2 =$ [ ]

$4 + 3 =$ [ ]

$8 \times 7 =$ [ ]

$15 \div 3 =$ [ ]

$6 + 1 =$ [ ]

$3 \times 5 =$ [ ]

$16 - 8 =$ [ ]

$14 - 9 =$ [ ]

$3 \times 6 =$ [ ]

$7 + 2 =$ [ ]

$7 \div 5 =$ [ ] REMAINDER [ ]

$2 + 4 =$ [ ]

$13 \div 5 =$ [ ] REMAINDER [ ]

End Time [ ] : [ ]

40

Day 13

Start Time ☐ : ☐

$5 - 0 =$ ☐

$3 + 3 =$ ☐

$9 \times 2 =$ ☐

$5 + 7 =$ ☐

$8 - 7 =$ ☐

$15 - 8 =$ ☐

$9 \div 9 =$ ☐

$9 \times 5 =$ ☐

$4 - 3 =$ ☐

$5 \div 2 =$ ☐ REMAINDER ☐

$7 \times 5 =$ ☐

$8 + 6 =$ ☐

$7 + 5 =$ ☐

$10 \div 8 =$ ☐ REMAINDER ☐

$6 \div 6 =$ ☐

$3 \times 0 =$ ☐

$8 \div 8 =$ ☐

$11 - 8 =$ ☐

$5 \times 4 =$ ☐

$11 - 4 =$ ☐

$9 + 6 =$ ☐

$4 \times 8 =$ ☐

$8 + 1 =$ ☐

$9 \div 3 =$ ☐

$8 \times 7 =$ ☐

$3 + 5 =$ ☐

$7 \div 1 =$ ☐

$18 - 9 =$ ☐

$7 - 5 =$ ☐

$6 + 3 =$ ☐

$3 - 2 =$ ☐

$9 \div 8 =$ ☐ REMAINDER ☐

$4 + 2 =$ ☐

$8 + 3 =$ ☐

$1 \times 6 =$ ☐

$14 - 6 =$ ☐

$9 \times 1 =$ ☐

$11 \div 4 =$ ☐ REMAINDER ☐

$7 \times 8 =$ ☐

$1 + 8 =$ ☐

$8 - 1 =$ ☐

$10 \div 4 =$ ☐ REMAINDER ☐

$15 - 9 =$ ☐

$12 \div 5 =$ ☐ REMAINDER ☐

$2 + 7 =$ ☐

$3 \times 1 =$ ☐

$8 \times 9 =$ ☐

$7 + 8 =$ ☐

$8 - 4 =$ ☐

$14 \div 7 =$ ☐

$9 \div 2 =$ ☐ REMAINDER ☐

$6 \times 9 =$ ☐

$3 \times 8 =$ ☐

$5 + 2 =$ ☐

$8 \div 4 =$ ☐

$10 \div 5 =$ ☐

$13 - 5 =$ ☐

$6 \times 6 =$ ☐

$10 - 4 =$ ☐

$15 \div 7 =$ ☐ REMAINDER ☐

$11 - 6 =$ ☐

$11 \div 7 =$ ☐ REMAINDER ☐

$9 \times 6 =$ ☐

$2 \times 9 =$ ☐

$3 \times 7 =$ ☐

$4 + 4 =$ ☐

$3 - 1 =$ ☐

$17 - 9 =$ ☐

$7 - 2 =$ ☐

$1 + 2 =$ ☐

$9 - 8 =$ ☐

$6 \times 5 =$ ☐

$18 \div 2 =$ ☐

$12 \div 6 =$ ☐

$7 + 3 =$ ☐

$8 - 3 =$ ☐

$6 + 1 =$ ☐

$4 \times 9 =$ ☐

$3 + 8 =$ ☐

$6 \times 2 =$ ☐

$14 - 5 =$ ☐

$6 + 5 =$ ☐

$8 + 5 =$ ☐

$7 \div 7 =$ ☐

$1 \times 9 =$ ☐

$10 - 2 =$ ☐

$3 \times 6 =$ ☐

$7 \div 4 =$ ☐ REMAINDER ☐

$6 + 7 =$ ☐

$4 + 3 =$ ☐

$2 + 6 =$ ☐

$5 + 5 =$ ☐

$5 \div 5 =$ ☐

$6 \div 5 =$ ☐ REMAINDER ☐

$6 \times 4 =$ ☐

$7 - 4 =$ ☐

$4 - 1 =$ ☐

$15 \div 3 =$ ☐

$18 \div 8 =$ ☐ REMAINDER ☐

$8 \times 6 =$ ☐

End Time ☐ : ☐

Date ☐ M ☐ D

Start Time ☐ : ☐

$7 - 7 =$ ☐

$4 + 9 =$ ☐

$18 \div 7 =$ ☐ REMAINDER ☐

$12 \div 9 =$ ☐ REMAINDER ☐

$1 \times 3 =$ ☐

$14 - 8 =$ ☐

$7 + 1 =$ ☐

$7 \times 4 =$ ☐

$2 + 5 =$ ☐

$8 - 2 =$ ☐

$2 - 1 =$ ☐

$5 \times 9 =$ ☐

$16 \div 4 =$ ☐

$15 \div 2 =$ ☐ REMAINDER ☐

$5 + 8 =$ ☐

$4 + 6 =$ ☐

$8 + 0 =$ ☐

$14 \div 2 =$ ☐

$8 \times 3 =$ ☐

$5 \div 3 =$ ☐ REMAINDER ☐

$9 \times 3 =$ ☐

$12 \div 3 =$ ☐

$9 + 4 =$ ☐

$3 - 3 =$ ☐

$4 + 2 =$ ☐

$7 - 3 =$ ☐

$4 + 1 =$ ☐

$9 - 4 =$ ☐

$4 \div 2 =$ ☐

$15 \div 5 =$ ☐

$3 \times 9 =$ ☐

$7 \times 6 =$ ☐

$1 \times 7 =$ ☐

$9 - 7 =$ ☐

$7 + 4 =$ ☐

$12 - 9 =$ ☐

$2 \times 7 =$ ☐

$3 + 9 =$ ☐

$7 \times 7 =$ ☐

$15 \div 4 =$ ☐ REMAINDER ☐

$10 - 5 =$ ☐

$3 + 4 =$ ☐

$14 - 7 =$ ☐

$15 - 7 =$ ☐

$13 - 4 =$ ☐

$3 + 6 =$ ☐

$4 \times 3 =$ ☐

$15 \div 6 =$ ☐ REMAINDER ☐

$7 \times 2 =$ ☐

$16 \div 8 =$ ☐

$4 - 4 =$ ⬚

$1 + 5 =$ ⬚

$15 - 6 =$ ⬚

$6 \div 4 =$ ⬚ REMAINDER ⬚

$8 \times 1 =$ ⬚

$6 + 6 =$ ⬚

$1 + 7 =$ ⬚

$10 \div 2 =$ ⬚

$4 \times 4 =$ ⬚

$13 - 8 =$ ⬚

$8 \times 8 =$ ⬚

$13 \div 6 =$ ⬚ REMAINDER ⬚

$8 - 6 =$ ⬚

$1 \times 5 =$ ⬚

$6 \div 2 =$ ⬚

$17 - 8 =$ ⬚

$3 + 2 =$ ⬚

$2 \times 3 =$ ⬚

$19 \div 4 =$ ⬚ REMAINDER ⬚

$5 \times 2 =$ ⬚

$7 + 7 =$ ⬚

$16 - 8 =$ ⬚

$10 - 9 =$ ⬚

$9 + 2 =$ ⬚

$5 \times 3 =$ ⬚

$18 \div 5 =$ ⬚ REMAINDER ⬚

$1 + 1 =$ ⬚

$9 \div 1 =$ ⬚

$2 + 2 =$ ⬚

$12 \div 2 =$ ⬚

$11 - 5 =$ ⬚

$5 \times 7 =$ ⬚

$3 \div 1 =$ ⬚

$4 \times 5 =$ ⬚

$6 \times 8 =$ ⬚

$4 + 7 =$ ⬚

$18 \div 6 =$ ⬚

$5 \times 8 =$ ⬚

$6 - 3 =$ ⬚

$18 \div 9 =$ ⬚

$1 + 9 =$ ⬚

$8 + 7 =$ ⬚

$9 - 2 =$ ⬚

$18 \div 4 =$ ⬚ REMAINDER ⬚

$3 \times 5 =$ ⬚

$7 - 2 =$ ⬚

$2 \times 6 =$ ⬚

$4 - 1 =$ ⬚

$6 + 2 =$ ⬚

$16 \div 6 =$ ⬚ REMAINDER ⬚

End Time ⬚ : ⬚

Date ☐ M ☐ D

Start Time ☐ : ☐

$11 - 2 =$ ☐

$1 + 5 =$ ☐

$2 \times 2 =$ ☐

$0 \times 4 =$ ☐

$13 \div 5 =$ ☐ REMAINDER ☐

$9 - 9 =$ ☐

$6 + 9 =$ ☐

$8 \div 8 =$ ☐

$4 + 6 =$ ☐

$8 - 6 =$ ☐

$4 \times 9 =$ ☐

$18 \div 3 =$ ☐

$4 - 3 =$ ☐

$3 + 5 =$ ☐

$2 \times 6 =$ ☐

$12 \div 7 =$ ☐ REMAINDER ☐

$10 \div 8 =$ ☐ REMAINDER ☐

$3 - 3 =$ ☐

$6 \times 1 =$ ☐

$5 + 6 =$ ☐

$3 + 8 =$ ☐

$14 - 5 =$ ☐

$3 + 2 =$ ☐

$0 \times 9 =$ ☐

$10 \div 2 =$ ☐

$9 \div 3 =$ ☐

$7 \times 8 =$ ☐

$12 - 9 =$ ☐

$6 \div 6 =$ ☐

$14 - 7 =$ ☐

$6 \times 5 =$ ☐

$7 + 3 =$ ☐

$4 \div 2 =$ ☐

$1 + 6 =$ ☐

$1 \times 3 =$ ☐

$2 - 1 =$ ☐

$10 \div 4 =$ ☐ REMAINDER ☐

$2 + 6 =$ ☐

$12 \div 9 =$ ☐ REMAINDER ☐

$3 \times 6 =$ ☐

$15 - 7 =$ ☐

$2 + 4 =$ ☐

$10 - 9 =$ ☐

$7 \times 2 =$ ☐

$6 - 3 =$ ☐

$11 \div 8 =$ ☐ REMAINDER ☐

$9 + 2 =$ ☐

$5 \times 4 =$ ☐

$9 - 2 =$ ☐

$7 + 1 =$ ☐

$9 \times 8 =$ [ ]

$14 \div 3 =$ [ ] REMAINDER [ ]

$15 \div 5 =$ [ ]

$9 - 5 =$ [ ]

$9 + 8 =$ [ ]

$5 \div 5 =$ [ ]

$3 \times 9 =$ [ ]

$2 + 2 =$ [ ]

$8 \div 6 =$ [ ] REMAINDER [ ]

$6 \times 6 =$ [ ]

$8 + 3 =$ [ ]

$7 - 7 =$ [ ]

$17 - 9 =$ [ ]

$2 + 1 =$ [ ]

$12 \div 3 =$ [ ]

$7 - 4 =$ [ ]

$4 \times 2 =$ [ ]

$5 \times 2 =$ [ ]

$8 \div 2 =$ [ ]

$8 - 3 =$ [ ]

$5 + 4 =$ [ ]

$15 - 6 =$ [ ]

$6 \div 2 =$ [ ]

$6 \times 9 =$ [ ]

$9 + 9 =$ [ ]

$9 - 1 =$ [ ]

$5 \times 1 =$ [ ]

$0 + 2 =$ [ ]

$3 \times 4 =$ [ ]

$7 \times 3 =$ [ ]

$13 - 4 =$ [ ]

$7 \div 1 =$ [ ]

$6 + 8 =$ [ ]

$10 \div 6 =$ [ ] REMAINDER [ ]

$1 + 3 =$ [ ]

$17 \div 7 =$ [ ] REMAINDER [ ]

$11 - 8 =$ [ ]

$8 \times 2 =$ [ ]

$3 + 3 =$ [ ]

$11 - 7 =$ [ ]

$5 + 9 =$ [ ]

$19 \div 2 =$ [ ] REMAINDER [ ]

$3 \times 8 =$ [ ]

$17 - 8 =$ [ ]

$9 \div 1 =$ [ ]

$8 \times 7 =$ [ ]

$4 + 8 =$ [ ]

$13 \div 7 =$ [ ] REMAINDER [ ]

$7 - 1 =$ [ ]

$2 \times 5 =$ [ ]

End Time [ ] : [ ]

I. Counting Test

Measure the time required for you to count from 1 to 120 aloud as fast as you can.

☐ sec.

II. Word Memorization Test

Memorize as many words as you can **within two minutes**.

smell	location	shade	eel	yore	offence
cigarette	vacuum	answer	tomorrow	freckle	mushroom
mussel	ally	movie	flounder	dawn	guidebook
baby	duster	raven	mole	caterpillar	heater
attitude	kimono	time	road	duration	carriage

Write out as many words as you can remember **in two minutes** on the back of this page. How many words can you remember?

Number of words memorized ☐ words

Word Memorization Test Answers

III. Stroop Test

Please take the **Stroop Test** for Week 3, located on page **vi** of the Appendix.

Start Time ☐ : ☐

2 × 9 = ☐

9 ÷ 4 = ☐ REMAINDER ☐

7 − 5 = ☐

4 + 9 = ☐

11 − 5 = ☐

7 × 3 = ☐

4 − 2 = ☐

8 × 9 = ☐

6 × 4 = ☐

8 ÷ 1 = ☐

7 + 7 = ☐

11 ÷ 7 = ☐ REMAINDER ☐

14 − 6 = ☐

2 + 3 = ☐

19 ÷ 5 = ☐ REMAINDER ☐

6 + 2 = ☐

8 − 7 = ☐

6 × 7 = ☐

10 ÷ 7 = ☐ REMAINDER ☐

5 + 1 = ☐

4 × 8 = ☐

2 + 1 = ☐

18 ÷ 5 = ☐ REMAINDER ☐

5 − 2 = ☐

8 + 6 = ☐

7 × 9 = ☐

8 ÷ 4 = ☐

3 + 4 = ☐

5 − 1 = ☐

3 + 7 = ☐

10 − 4 = ☐

8 × 8 = ☐

12 ÷ 6 = ☐

8 × 5 = ☐

2 + 7 = ☐

8 × 4 = ☐

14 ÷ 2 = ☐

8 + 7 = ☐

10 − 6 = ☐

16 ÷ 8 = ☐

7 − 3 = ☐

7 ÷ 2 = ☐ REMAINDER ☐

13 − 9 = ☐

10 − 5 = ☐

5 + 3 = ☐

7 × 7 = ☐

15 ÷ 3 = ☐

9 × 4 = ☐

2 − 0 = ☐

9 + 1 = ☐

$10 \div 5 =$ [ ]

$2 + 9 =$ [ ]

$5 \times 7 =$ [ ]

$15 - 8 =$ [ ]

$9 \div 9 =$ [ ]

$1 \times 4 =$ [ ]

$12 - 5 =$ [ ]

$4 + 2 =$ [ ]

$1 + 4 =$ [ ]

$10 - 7 =$ [ ]

$2 \times 8 =$ [ ]

$18 \div 2 =$ [ ]

$6 \times 3 =$ [ ]

$7 + 8 =$ [ ]

$6 - 1 =$ [ ]

$4 \times 7 =$ [ ]

$6 \div 4 =$ [ ] REMAINDER [ ]

$15 \div 8 =$ [ ] REMAINDER [ ]

$4 \times 5 =$ [ ]

$6 + 3 =$ [ ]

$7 - 0 =$ [ ]

$4 \div 3 =$ [ ] REMAINDER [ ]

$9 + 7 =$ [ ]

$9 \times 3 =$ [ ]

$7 - 1 =$ [ ]

$16 \div 9 =$ [ ] REMAINDER [ ]

$6 - 2 =$ [ ]

$5 + 7 =$ [ ]

$12 \div 4 =$ [ ]

$3 \times 5 =$ [ ]

$2 \times 4 =$ [ ]

$10 - 1 =$ [ ]

$7 + 2 =$ [ ]

$5 \div 1 =$ [ ]

$5 + 2 =$ [ ]

$7 \div 4 =$ [ ] REMAINDER [ ]

$11 - 4 =$ [ ]

$3 \times 2 =$ [ ]

$1 + 8 =$ [ ]

$16 \div 5 =$ [ ] REMAINDER [ ]

$14 - 9 =$ [ ]

$9 \times 2 =$ [ ]

$5 + 5 =$ [ ]

$6 \div 3 =$ [ ]

$6 + 7 =$ [ ]

$4 \times 3 =$ [ ]

$8 - 4 =$ [ ]

$12 \div 2 =$ [ ]

$6 \times 2 =$ [ ]

$9 - 8 =$ [ ]

End Time [ ] : [ ]

Start Time ☐ : ☐

13 − 4 = ☐

10 ÷ 5 = ☐

8 × 6 = ☐

9 − 0 = ☐

1 + 6 = ☐

8 × 7 = ☐

16 ÷ 4 = ☐

1 + 3 = ☐

14 ÷ 7 = ☐

8 − 1 = ☐

7 × 6 = ☐

8 + 3 = ☐

9 − 9 = ☐

12 ÷ 4 = ☐

9 + 3 = ☐

0 × 7 = ☐

8 ÷ 5 = ☐ REMAINDER ☐

9 − 2 = ☐

1 × 9 = ☐

7 × 4 = ☐

8 − 3 = ☐

3 + 6 = ☐

15 ÷ 6 = ☐ REMAINDER ☐

14 ÷ 4 = ☐ REMAINDER ☐

9 + 5 = ☐

2 × 7 = ☐

18 ÷ 9 = ☐

6 × 6 = ☐

11 − 9 = ☐

2 + 6 = ☐

10 − 5 = ☐

11 ÷ 6 = ☐ REMAINDER ☐

9 + 2 = ☐

6 − 5 = ☐

2 × 6 = ☐

12 − 3 = ☐

3 + 3 = ☐

10 ÷ 3 = ☐ REMAINDER ☐

5 × 6 = ☐

14 − 7 = ☐

8 − 5 = ☐

6 + 3 = ☐

4 × 5 = ☐

4 ÷ 2 = ☐

8 + 6 = ☐

11 − 5 = ☐

7 × 7 = ☐

4 + 9 = ☐

11 ÷ 4 = ☐ REMAINDER ☐

8 + 1 = ☐

$13 - 6 =$ ☐

$9 \times 4 =$ ☐

$6 \times 2 =$ ☐

$11 - 7 =$ ☐

$8 + 9 =$ ☐

$8 \div 2 =$ ☐

$2 + 1 =$ ☐

$7 \div 2 =$ ☐ REMAINDER ☐

$9 \times 2 =$ ☐

$6 \div 3 =$ ☐

$4 - 2 =$ ☐

$2 \times 9 =$ ☐

$4 \div 1 =$ ☐

$6 + 5 =$ ☐

$8 \times 5 =$ ☐

$10 - 4 =$ ☐

$4 + 6 =$ ☐

$11 \div 3 =$ ☐ REMAINDER ☐

$3 + 2 =$ ☐

$5 \times 2 =$ ☐

$8 \div 8 =$ ☐

$3 - 1 =$ ☐

$4 \times 6 =$ ☐

$1 \times 7 =$ ☐

$12 - 9 =$ ☐

$2 \div 2 =$ ☐

$8 \div 3 =$ ☐ REMAINDER ☐

$7 - 5 =$ ☐

$3 + 5 =$ ☐

$6 + 9 =$ ☐

$16 \div 8 =$ ☐

$2 \times 2 =$ ☐

$8 - 4 =$ ☐

$2 + 5 =$ ☐

$15 - 8 =$ ☐

$9 - 4 =$ ☐

$7 + 2 =$ ☐

$1 \times 6 =$ ☐

$19 \div 2 =$ ☐ REMAINDER ☐

$5 \div 4 =$ ☐ REMAINDER ☐

$7 \times 3 =$ ☐

$5 - 4 =$ ☐

$4 + 2 =$ ☐

$7 \div 1 =$ ☐

$9 \times 8 =$ ☐

$9 + 6 =$ ☐

$17 \div 5 =$ ☐ REMAINDER ☐

$6 \times 5 =$ ☐

$15 - 9 =$ ☐

$5 + 7 =$ ☐

End Time ☐ : ☐

Date ☐ M ☐ D

Start Time ☐ : ☐

$9 - 8 =$ ☐

$14 \div 3 =$ ☐ REMAINDER ☐

$9 \times 6 =$ ☐

$2 + 4 =$ ☐

$15 \div 7 =$ ☐ REMAINDER ☐

$6 \times 8 =$ ☐

$6 \div 2 =$ ☐

$7 \times 2 =$ ☐

$8 - 6 =$ ☐

$1 + 5 =$ ☐

$5 + 3 =$ ☐

$12 \div 9 =$ ☐ REMAINDER ☐

$6 - 1 =$ ☐

$2 \times 1 =$ ☐

$1 + 7 =$ ☐

$2 - 2 =$ ☐

$4 \times 8 =$ ☐

$3 \div 3 =$ ☐

$9 + 9 =$ ☐

$9 - 3 =$ ☐

$3 \times 2 =$ ☐

$7 + 9 =$ ☐

$17 - 8 =$ ☐

$2 + 2 =$ ☐

$11 \div 9 =$ ☐ REMAINDER ☐

$3 \times 9 =$ ☐

$2 + 3 =$ ☐

$9 - 5 =$ ☐

$12 \div 2 =$ ☐

$6 \times 9 =$ ☐

$12 \div 3 =$ ☐

$14 - 6 =$ ☐

$7 + 6 =$ ☐

$15 \div 3 =$ ☐

$0 \times 5 =$ ☐

$12 - 8 =$ ☐

$14 - 5 =$ ☐

$5 + 4 =$ ☐

$17 - 9 =$ ☐

$4 \times 2 =$ ☐

$18 \div 3 =$ ☐

$13 - 7 =$ ☐

$6 + 8 =$ ☐

$5 - 3 =$ ☐

$5 \times 1 =$ ☐

$11 \div 2 =$ ☐ REMAINDER ☐

$2 + 9 =$ ☐

$1 \times 8 =$ ☐

$17 \div 7 =$ ☐ REMAINDER ☐

$8 + 8 =$ ☐

$1 \times 3 =$ [   ] $11 - 8 =$ [   ] $5 + 2 =$ [   ]

$9 + 7 =$ [   ] $8 \times 3 =$ [   ] $13 \div 3 =$ [   ] REMAINDER [   ]

$11 - 3 =$ [   ] $1 + 1 =$ [   ] $11 - 2 =$ [   ]

$9 + 4 =$ [   ] $4 - 3 =$ [   ] $7 \times 8 =$ [   ]

$8 \times 8 =$ [   ] $11 \div 8 =$ [   ] REMAINDER [   ] $0 + 3 =$ [   ]

$16 \div 2 =$ [   ] $1 + 2 =$ [   ] $6 \times 1 =$ [   ]

$3 + 4 =$ [   ] $15 \div 5 =$ [   ] $4 + 4 =$ [   ]

$19 \div 8 =$ [   ] REMAINDER [   ] $3 \times 3 =$ [   ] $2 \div 1 =$ [   ]

$5 - 4 =$ [   ] $7 + 5 =$ [   ] $4 \times 7 =$ [   ]

$8 \times 4 =$ [   ] $9 \times 7 =$ [   ] $11 \div 5 =$ [   ] REMAINDER [   ]

$14 - 8 =$ [   ] $10 - 6 =$ [   ] $7 - 2 =$ [   ]

$3 \times 4 =$ [   ] $10 \div 2 =$ [   ] $3 + 8 =$ [   ]

$12 \div 6 =$ [   ] $7 \times 5 =$ [   ] $6 \times 3 =$ [   ]

$19 \div 4 =$ [   ] REMAINDER [   ] $4 - 2 =$ [   ] $3 - 2 =$ [   ]

$7 + 8 =$ [   ] $12 \div 5 =$ [   ] REMAINDER [   ] $18 \div 2 =$ [   ]

$8 - 8 =$ [   ] $3 \times 6 =$ [   ] $13 - 8 =$ [   ]

$8 \div 4 =$ [   ] $8 + 2 =$ [   ]   **End Time** [   ] : [   ]

Day 19

Start Time ☐ : ☐

$4 + 5 =$ ☐

$15 - 9 =$ ☐

$2 \times 3 =$ ☐

$7 \div 3 =$ ☐ REMAINDER ☐

$0 \times 5 =$ ☐

$7 + 9 =$ ☐

$14 \div 2 =$ ☐

$6 - 1 =$ ☐

$2 \times 2 =$ ☐

$6 + 2 =$ ☐

$8 + 6 =$ ☐

$16 \div 3 =$ ☐ REMAINDER ☐

$9 \times 5 =$ ☐

$7 \div 1 =$ ☐

$7 - 3 =$ ☐

$8 - 5 =$ ☐

$14 \div 4 =$ ☐ REMAINDER ☐

$5 + 6 =$ ☐

$16 - 9 =$ ☐

$12 \div 9 =$ ☐ REMAINDER ☐

$8 - 6 =$ ☐

$5 \times 2 =$ ☐

$1 + 7 =$ ☐

$17 - 8 =$ ☐

$7 \div 4 =$ ☐ REMAINDER ☐

$2 + 2 =$ ☐

$6 \div 6 =$ ☐

$7 \times 7 =$ ☐

$4 - 3 =$ ☐

$8 \times 6 =$ ☐

$14 - 7 =$ ☐

$8 + 2 =$ ☐

$2 \times 4 =$ ☐

$8 + 7 =$ ☐

$9 \times 4 =$ ☐

$10 - 6 =$ ☐

$7 + 2 =$ ☐

$4 \times 2 =$ ☐

$6 \div 5 =$ ☐ REMAINDER ☐

$8 \times 7 =$ ☐

$6 - 5 =$ ☐

$12 \div 2 =$ ☐

$7 \times 6 =$ ☐

$5 + 2 =$ ☐

$8 - 1 =$ ☐

$12 \div 3 =$ ☐

$6 + 6 =$ ☐

$10 - 5 =$ ☐

$15 \div 5 =$ ☐

$1 + 1 =$ ☐

$11 - 3 =$ [ ]

$7 \div 5 =$ [ ] REMAINDER [ ]

$5 \div 4 =$ [ ] REMAINDER [ ]

$14 \div 9 =$ [ ] REMAINDER [ ]

$5 + 4 =$ [ ]

$2 + 9 =$ [ ]

$15 \div 2 =$ [ ] REMAINDER [ ]

$4 - 1 =$ [ ]

$9 - 6 =$ [ ]

$5 + 1 =$ [ ]

$1 \times 7 =$ [ ]

$8 + 1 =$ [ ]

$9 \times 6 =$ [ ]

$9 - 5 =$ [ ]

$12 \div 4 =$ [ ]

$14 - 9 =$ [ ]

$2 + 8 =$ [ ]

$3 \times 5 =$ [ ]

$4 \div 4 =$ [ ]

$6 \times 3 =$ [ ]

$9 + 8 =$ [ ]

$5 \times 0 =$ [ ]

$10 - 7 =$ [ ]

$1 + 3 =$ [ ]

$9 \div 8 =$ [ ] REMAINDER [ ]

$9 \times 2 =$ [ ]

$5 \times 5 =$ [ ]

$5 + 3 =$ [ ]

$5 \times 1 =$ [ ]

$5 - 4 =$ [ ]

$2 \times 6 =$ [ ]

$15 \div 8 =$ [ ] REMAINDER [ ]

$14 \div 7 =$ [ ]

$9 + 2 =$ [ ]

$16 - 7 =$ [ ]

$6 \times 2 =$ [ ]

$15 \div 3 =$ [ ]

$16 \div 2 =$ [ ]

$17 - 9 =$ [ ]

$8 - 2 =$ [ ]

$7 + 0 =$ [ ]

$12 \div 6 =$ [ ]

$7 \times 4 =$ [ ]

$5 + 8 =$ [ ]

$7 \times 2 =$ [ ]

$9 + 9 =$ [ ]

$2 \times 1 =$ [ ]

$13 - 9 =$ [ ]

$3 - 1 =$ [ ]

$18 \div 9 =$ [ ]

End Time [ ] : [ ]

Day 20

Start Time ☐ : ☐

$13 - 8 =$ ☐

$8 + 5 =$ ☐

$12 - 7 =$ ☐

$8 \div 3 =$ ☐ REMAINDER ☐

$7 \times 5 =$ ☐

$4 + 2 =$ ☐

$3 \times 9 =$ ☐

$6 - 4 =$ ☐

$1 + 3 =$ ☐

$9 \div 5 =$ ☐ REMAINDER ☐

$4 \times 4 =$ ☐

$7 - 6 =$ ☐

$2 \div 2 =$ ☐

$6 + 3 =$ ☐

$5 \times 9 =$ ☐

$5 \div 5 =$ ☐

$17 \div 6 =$ ☐ REMAINDER ☐

$9 + 5 =$ ☐

$7 - 4 =$ ☐

$6 + 7 =$ ☐

$8 \div 4 =$ ☐

$9 \times 9 =$ ☐

$2 \times 8 =$ ☐

$9 + 7 =$ ☐

$10 \div 2 =$ ☐

$9 - 9 =$ ☐

$6 \times 7 =$ ☐

$2 + 5 =$ ☐

$8 \times 3 =$ ☐

$17 \div 3 =$ ☐ REMAINDER ☐

$2 \times 7 =$ ☐

$1 + 5 =$ ☐

$10 - 9 =$ ☐

$16 \div 8 =$ ☐

$9 - 3 =$ ☐

$1 + 2 =$ ☐

$11 - 6 =$ ☐

$9 - 4 =$ ☐

$11 \div 4 =$ ☐ REMAINDER ☐

$12 - 8 =$ ☐

$3 + 3 =$ ☐

$4 + 7 =$ ☐

$8 - 0 =$ ☐

$16 \div 4 =$ ☐

$6 \times 4 =$ ☐

$16 - 8 =$ ☐

$6 + 5 =$ ☐

$4 \times 7 =$ ☐

$15 \div 6 =$ ☐ REMAINDER ☐

$4 \times 8 =$ ☐

$8 \times 9 =$ [ ]

$8 + 4 =$ [ ]

$18 - 9 =$ [ ]

$13 \div 2 =$ [ ] REMAINDER [ ]

$1 \times 5 =$ [ ]

$13 - 4 =$ [ ]

$5 \times 4 =$ [ ]

$7 - 1 =$ [ ]

$8 \div 2 =$ [ ]

$6 - 2 =$ [ ]

$2 \times 9 =$ [ ]

$18 \div 8 =$ [ ] REMAINDER [ ]

$4 + 6 =$ [ ]

$3 + 4 =$ [ ]

$6 \div 2 =$ [ ]

$7 \times 0 =$ [ ]

$8 + 8 =$ [ ]

$14 \div 6 =$ [ ] REMAINDER [ ]

$3 \times 6 =$ [ ]

$10 \div 5 =$ [ ]

$3 - 1 =$ [ ]

$6 \div 3 =$ [ ]

$3 + 6 =$ [ ]

$13 - 7 =$ [ ]

$3 \times 8 =$ [ ]

$8 \div 6 =$ [ ] REMAINDER [ ]

$2 + 6 =$ [ ]

$9 - 2 =$ [ ]

$3 + 2 =$ [ ]

$9 \div 9 =$ [ ]

$4 \times 3 =$ [ ]

$9 \times 7 =$ [ ]

$8 - 2 =$ [ ]

$3 + 7 =$ [ ]

$6 \times 8 =$ [ ]

$6 - 6 =$ [ ]

$9 \div 4 =$ [ ] REMAINDER [ ]

$9 \times 1 =$ [ ]

$15 - 7 =$ [ ]

$4 + 8 =$ [ ]

$12 - 6 =$ [ ]

$9 + 1 =$ [ ]

$6 \times 6 =$ [ ]

$1 + 6 =$ [ ]

$18 \div 3 =$ [ ]

$0 + 2 =$ [ ]

$15 \div 4 =$ [ ] REMAINDER [ ]

$8 \times 5 =$ [ ]

$11 - 4 =$ [ ]

$18 \div 6 =$ [ ]

End Time [ ] : [ ]

I. Counting Test

Measure the time required for you to count from 1 to 120 aloud as fast as you can.

☐ sec.

II. Word Memorization Test

Memorize as many words as you can **within two minutes**.

servant	temple	danger	tiptoe	stove	kangaroo
oval	knowledge	lifetime	scissor	counter	packing
damson	exit	reed	apple	blanket	broom
epoch	south	order	sardine	corridor	hailstone
courtesy	earthquake	forehead	autumn	sequel	action

Write out as many words as you can remember **in two minutes** on the back of this page. How many words can you remember?

Number of words memorized ☐ words

Word Memorization Test Answers

III. Stroop Test

Please take the **Stroop Test** for Week 4, located on page **vii** of the Appendix.

Start Time ☐ : ☐

$2 + 1 =$ ☐

$18 \div 5 =$ ☐ REMAINDER ☐

$5 \times 2 =$ ☐

$17 - 9 =$ ☐

$1 + 4 =$ ☐

$7 + 7 =$ ☐

$14 \div 7 =$ ☐

$9 \times 6 =$ ☐

$10 - 8 =$ ☐

$18 \div 6 =$ ☐

$10 - 9 =$ ☐

$6 \times 5 =$ ☐

$13 \div 7 =$ ☐ REMAINDER ☐

$4 \times 1 =$ ☐

$3 - 1 =$ ☐

$1 + 6 =$ ☐

$12 \div 4 =$ ☐

$4 \times 2 =$ ☐

$1 \times 1 =$ ☐

$7 + 9 =$ ☐

$5 + 1 =$ ☐

$4 - 3 =$ ☐

$3 - 0 =$ ☐

$5 \div 1 =$ ☐

$2 \times 6 =$ ☐

$6 - 5 =$ ☐

$2 + 7 =$ ☐

$13 - 9 =$ ☐

$9 + 9 =$ ☐

$15 \div 9 =$ ☐ REMAINDER ☐

$17 \div 8 =$ ☐ REMAINDER ☐

$7 \times 0 =$ ☐

$5 + 6 =$ ☐

$3 + 8 =$ ☐

$8 \div 4 =$ ☐

$1 \times 7 =$ ☐

$18 - 9 =$ ☐

$1 + 5 =$ ☐

$9 - 2 =$ ☐

$7 \times 6 =$ ☐

$14 \div 4 =$ ☐ REMAINDER ☐

$4 - 2 =$ ☐

$7 \times 7 =$ ☐

$5 + 2 =$ ☐

$3 \times 7 =$ ☐

$9 + 3 =$ ☐

$15 \div 3 =$ ☐

$11 - 8 =$ ☐

$13 \div 9 =$ ☐ REMAINDER ☐

$9 - 6 =$ ☐

Time Required [   ] : [   ]

$9 \div 2 =$ [   ] REMAINDER [   ]

$12 \div 2 =$ [   ]

$5 + 9 =$ [   ]

$8 - 7 =$ [   ]

$7 \times 5 =$ [   ]

$12 - 3 =$ [   ]

$8 \div 8 =$ [   ]

$8 + 9 =$ [   ]

$2 \times 2 =$ [   ]

$8 + 0 =$ [   ]

$18 \div 2 =$ [   ]

$12 - 9 =$ [   ]

$4 \times 9 =$ [   ]

$4 \div 2 =$ [   ]

$7 + 2 =$ [   ]

$10 - 6 =$ [   ]

$4 \times 6 =$ [   ]

$8 \times 9 =$ [   ]

$13 - 7 =$ [   ]

$9 \div 4 =$ [   ] REMAINDER [   ]

$7 + 1 =$ [   ]

$9 \div 1 =$ [   ]

$7 - 6 =$ [   ]

$5 - 1 =$ [   ]

$2 + 9 =$ [   ]

$11 \div 3 =$ [   ] REMAINDER [   ]

$8 \times 4 =$ [   ]

$5 + 3 =$ [   ]

$7 \times 9 =$ [   ]

$11 - 2 =$ [   ]

$2 \times 9 =$ [   ]

$8 + 6 =$ [   ]

$6 \div 2 =$ [   ]

$4 \times 7 =$ [   ]

$8 - 1 =$ [   ]

$8 \times 5 =$ [   ]

$1 + 1 =$ [   ]

$19 \div 5 =$ [   ] REMAINDER [   ]

$7 + 5 =$ [   ]

$9 \times 8 =$ [   ]

$3 \times 4 =$ [   ]

$16 \div 2 =$ [   ]

$6 - 2 =$ [   ]

$15 \div 8 =$ [   ] REMAINDER [   ]

$5 - 3 =$ [   ]

$4 + 9 =$ [   ]

$2 + 3 =$ [   ]

$17 \div 6 =$ [   ] REMAINDER [   ]

$6 \times 6 =$ [   ]

$13 - 8 =$ [   ]

End Time [   ] : [   ]

Date ☐ M ☐ D

Start Time ☐ : ☐

6 + 1 = ☐

15 − 8 = ☐

7 ÷ 7 = ☐

8 × 2 = ☐

6 + 7 = ☐

7 − 3 = ☐

3 + 5 = ☐

9 × 9 = ☐

6 ÷ 6 = ☐

6 × 9 = ☐

14 ÷ 8 = ☐ REMAINDER ☐

7 − 2 = ☐

9 × 3 = ☐

4 + 7 = ☐

18 ÷ 8 = ☐ REMAINDER ☐

11 − 4 = ☐

3 + 2 = ☐

9 ÷ 3 = ☐

5 × 5 = ☐

9 − 8 = ☐

19 ÷ 4 = ☐ REMAINDER ☐

9 + 6 = ☐

4 × 8 = ☐

7 − 1 = ☐

1 + 8 = ☐

4 − 1 = ☐

2 × 4 = ☐

7 − 5 = ☐

9 ÷ 6 = ☐ REMAINDER ☐

6 − 3 = ☐

9 × 4 = ☐

12 ÷ 6 = ☐

6 + 6 = ☐

16 ÷ 3 = ☐ REMAINDER ☐

1 + 2 = ☐

12 − 8 = ☐

18 ÷ 3 = ☐

3 × 5 = ☐

16 − 9 = ☐

12 − 4 = ☐

1 × 9 = ☐

7 × 8 = ☐

9 + 1 = ☐

16 ÷ 8 = ☐

6 + 8 = ☐

4 + 1 = ☐

8 × 6 = ☐

14 ÷ 3 = ☐ REMAINDER ☐

12 − 5 = ☐

6 + 3 = ☐

$6 + 4 =$ ☐

$9 \div 9 =$ ☐

$10 \div 6 =$ ☐ REMAINDER ☐

$3 \times 1 =$ ☐

$3 + 7 =$ ☐

$6 \times 4 =$ ☐

$5 - 2 =$ ☐

$17 \div 3 =$ ☐ REMAINDER ☐

$5 \times 9 =$ ☐

$16 \div 5 =$ ☐ REMAINDER ☐

$8 + 1 =$ ☐

$8 - 4 =$ ☐

$7 \div 2 =$ ☐ REMAINDER ☐

$12 - 7 =$ ☐

$8 + 5 =$ ☐

$1 \times 4 =$ ☐

$7 - 4 =$ ☐

$8 \times 7 =$ ☐

$2 \times 3 =$ ☐

$1 + 3 =$ ☐

$6 \div 3 =$ ☐

$11 - 5 =$ ☐

$10 \div 2 =$ ☐

$5 \times 0 =$ ☐

$11 - 7 =$ ☐

$4 + 2 =$ ☐

$7 - 6 =$ ☐

$9 + 7 =$ ☐

$14 \div 2 =$ ☐

$4 + 4 =$ ☐

$8 \times 3 =$ ☐

$12 \div 3 =$ ☐

$15 - 7 =$ ☐

$3 \times 6 =$ ☐

$7 \times 2 =$ ☐

$3 + 4 =$ ☐

$2 - 2 =$ ☐

$12 \div 2 =$ ☐

$6 \times 7 =$ ☐

$2 \times 8 =$ ☐

$10 \div 7 =$ ☐ REMAINDER ☐

$6 - 4 =$ ☐

$15 \div 5 =$ ☐

$5 \times 3 =$ ☐

$2 + 2 =$ ☐

$7 \div 3 =$ ☐ REMAINDER ☐

$13 - 6 =$ ☐

$7 + 3 =$ ☐

$14 - 9 =$ ☐

$5 + 5 =$ ☐

End Time ☐ : ☐

64

Date ☐ M ☐ D

Start Time ☐ : ☐

$7 \div 1 =$ ☐

$8 \times 0 =$ ☐

$17 - 9 =$ ☐

$6 + 3 =$ ☐

$11 \div 9 =$ ☐ REMAINDER ☐

$3 \times 9 =$ ☐

$1 + 6 =$ ☐

$8 - 0 =$ ☐

$2 \times 3 =$ ☐

$3 + 8 =$ ☐

$2 + 6 =$ ☐

$1 \div 1 =$ ☐

$5 \times 6 =$ ☐

$6 - 5 =$ ☐

$7 - 3 =$ ☐

$16 \div 4 =$ ☐

$4 \times 1 =$ ☐

$5 + 4 =$ ☐

$8 \div 2 =$ ☐

$6 \times 3 =$ ☐

$17 \div 5 =$ ☐ REMAINDER ☐

$2 + 9 =$ ☐

$14 - 7 =$ ☐

$3 + 5 =$ ☐

$5 \times 7 =$ ☐

$7 - 6 =$ ☐

$12 \div 9 =$ ☐ REMAINDER ☐

$6 - 1 =$ ☐

$3 + 6 =$ ☐

$11 - 4 =$ ☐

$14 \div 7 =$ ☐

$9 + 1 =$ ☐

$7 \times 0 =$ ☐

$12 \div 7 =$ ☐ REMAINDER ☐

$3 \times 2 =$ ☐

$6 \div 4 =$ ☐ REMAINDER ☐

$12 - 3 =$ ☐

$8 \times 9 =$ ☐

$5 + 1 =$ ☐

$5 \times 5 =$ ☐

$7 \times 4 =$ ☐

$9 + 5 =$ ☐

$6 - 3 =$ ☐

$12 - 4 =$ ☐

$6 \div 3 =$ ☐

$9 + 3 =$ ☐

$5 \div 4 =$ ☐ REMAINDER ☐

$10 - 3 =$ ☐

$9 - 0 =$ ☐

$7 + 9 =$ ☐

$2 + 1 =$ ☐

$1 \times 4 =$ ☐

$11 - 6 =$ ☐

$6 + 8 =$ ☐

$9 \times 6 =$ ☐

$7 - 5 =$ ☐

$15 \div 3 =$ ☐

$3 + 7 =$ ☐

$10 \div 9 =$ ☐ REMAINDER ☐

$1 \times 2 =$ ☐

$11 - 9 =$ ☐

$3 \div 3 =$ ☐

$8 \times 5 =$ ☐

$5 \div 2 =$ ☐ REMAINDER ☐

$7 + 1 =$ ☐

$10 - 4 =$ ☐

$2 \div 2 =$ ☐

$9 \times 9 =$ ☐

$8 - 3 =$ ☐

$9 \div 6 =$ ☐ REMAINDER ☐

$6 \times 7 =$ ☐

$5 + 6 =$ ☐

$12 \div 2 =$ ☐

$13 - 7 =$ ☐

$7 + 4 =$ ☐

$1 + 3 =$ ☐

$5 \times 4 =$ ☐

$12 - 8 =$ ☐

$18 \div 6 =$ ☐

$3 + 4 =$ ☐

$5 \times 3 =$ ☐

$4 \times 5 =$ ☐

$18 \div 8 =$ ☐ REMAINDER ☐

$5 + 9 =$ ☐

$2 - 2 =$ ☐

$13 \div 7 =$ ☐ REMAINDER ☐

$8 - 4 =$ ☐

$1 + 5 =$ ☐

$10 - 9 =$ ☐

$4 \div 1 =$ ☐

$5 \times 9 =$ ☐

$1 + 4 =$ ☐

$6 - 2 =$ ☐

$7 \times 5 =$ ☐

$1 \times 9 =$ ☐

$19 \div 7 =$ ☐ REMAINDER ☐

$3 - 2 =$ ☐

$10 \div 2 =$ ☐

$8 + 6 =$ ☐

$9 \times 7 =$ ☐

End Time ☐ : ☐

Start Time ☐ : ☐

$2 \times 5 =$ ☐

$10 - 1 =$ ☐

$9 - 9 =$ ☐

$18 \div 2 =$ ☐

$2 - 1 =$ ☐

$9 \times 2 =$ ☐

$6 + 1 =$ ☐

$7 \div 2 =$ ☐ REMAINDER ☐

$4 + 4 =$ ☐

$6 \div 2 =$ ☐

$6 - 6 =$ ☐

$4 \times 4 =$ ☐

$15 \div 7 =$ ☐ REMAINDER ☐

$6 + 7 =$ ☐

$1 + 3 =$ ☐

$9 \times 1 =$ ☐

$13 \div 4 =$ ☐ REMAINDER ☐

$5 + 5 =$ ☐

$6 \times 8 =$ ☐

$8 - 1 =$ ☐

$6 \times 6 =$ ☐

$16 \div 7 =$ ☐ REMAINDER ☐

$4 + 1 =$ ☐

$16 - 7 =$ ☐

$8 + 2 =$ ☐

$2 \div 1 =$ ☐

$4 + 3 =$ ☐

$3 \times 8 =$ ☐

$17 \div 4 =$ ☐ REMAINDER ☐

$11 - 5 =$ ☐

$6 \times 5 =$ ☐

$3 + 0 =$ ☐

$12 - 9 =$ ☐

$4 \times 6 =$ ☐

$15 - 8 =$ ☐

$9 + 7 =$ ☐

$9 \times 8 =$ ☐

$4 + 6 =$ ☐

$4 - 2 =$ ☐

$9 - 4 =$ ☐

$2 \times 4 =$ ☐

$6 \div 6 =$ ☐

$10 - 2 =$ ☐

$8 \div 4 =$ ☐

$5 + 2 =$ ☐

$3 + 9 =$ ☐

$9 \div 8 =$ ☐ REMAINDER ☐

$5 - 3 =$ ☐

$9 \times 5 =$ ☐

$16 \div 8 =$ ☐

$9 + 2 =$ ☐

$10 - 7 =$ ☐

$3 \div 1 =$ ☐

$6 \times 4 =$ ☐

$4 + 5 =$ ☐

$12 \div 2 =$ ☐

$16 - 8 =$ ☐

$2 \times 6 =$ ☐

$12 \div 4 =$ ☐

$1 \times 8 =$ ☐

$14 - 8 =$ ☐

$8 + 7 =$ ☐

$6 \div 1 =$ ☐

$3 \times 6 =$ ☐

$1 \times 3 =$ ☐

$6 + 9 =$ ☐

$11 \div 8 =$ ☐ REMAINDER ☐

$4 \times 7 =$ ☐

$16 \div 2 =$ ☐

$12 - 5 =$ ☐

$9 \times 3 =$ ☐

$17 \div 7 =$ ☐ REMAINDER ☐

$1 + 1 =$ ☐

$12 \div 5 =$ ☐ REMAINDER ☐

$2 \times 8 =$ ☐

$8 - 8 =$ ☐

$7 + 5 =$ ☐

$4 - 3 =$ ☐

$9 - 5 =$ ☐

$7 \times 7 =$ ☐

$6 + 2 =$ ☐

$4 + 7 =$ ☐

$15 \div 8 =$ ☐ REMAINDER ☐

$9 - 3 =$ ☐

$7 + 2 =$ ☐

$18 \div 9 =$ ☐

$1 + 2 =$ ☐

$4 + 8 =$ ☐

$8 \times 7 =$ ☐

$5 \div 3 =$ ☐ REMAINDER ☐

$3 \times 5 =$ ☐

$18 - 9 =$ ☐

$2 - 2 =$ ☐

$1 \times 1 =$ ☐

$11 - 8 =$ ☐

$3 \div 2 =$ ☐ REMAINDER ☐

$2 \times 9 =$ ☐

$9 \div 3 =$ ☐

$3 + 3 =$ ☐

$5 - 2 =$ ☐

End Time ☐ : ☐

Date ☐ M ☐ D

Start Time ☐ : ☐

$3 \times 0 =$ ☐

$12 - 8 =$ ☐

$9 \div 2 =$ ☐ REMAINDER ☐

$6 + 9 =$ ☐

$2 + 7 =$ ☐

$5 - 4 =$ ☐

$5 \times 2 =$ ☐

$12 \div 6 =$ ☐

$4 + 8 =$ ☐

$8 \div 8 =$ ☐

$9 \times 3 =$ ☐

$0 \times 9 =$ ☐

$13 - 6 =$ ☐

$10 - 1 =$ ☐

$17 \div 8 =$ ☐ REMAINDER ☐

$1 + 5 =$ ☐

$2 - 1 =$ ☐

$3 \times 7 =$ ☐

$9 + 5 =$ ☐

$14 \div 9 =$ ☐ REMAINDER ☐

$4 \times 8 =$ ☐

$10 - 2 =$ ☐

$9 \times 6 =$ ☐

$8 + 0 =$ ☐

$9 \div 9 =$ ☐

$7 + 1 =$ ☐

$9 - 1 =$ ☐

$6 \div 3 =$ ☐

$7 - 2 =$ ☐

$15 \div 6 =$ ☐ REMAINDER ☐

$1 + 8 =$ ☐

$7 - 0 =$ ☐

$1 \times 7 =$ ☐

$15 \div 9 =$ ☐ REMAINDER ☐

$1 + 4 =$ ☐

$12 \div 3 =$ ☐

$9 + 8 =$ ☐

$5 \times 0 =$ ☐

$11 - 9 =$ ☐

$10 \div 5 =$ ☐

$6 + 4 =$ ☐

$9 + 0 =$ ☐

$19 \div 2 =$ ☐ REMAINDER ☐

$8 \times 0 =$ ☐

$1 \times 8 =$ ☐

$5 - 3 =$ ☐

$10 - 5 =$ ☐

$8 \times 4 =$ ☐

$1 + 9 =$ ☐

$6 - 4 =$ ☐

$15 \div 8 =$ [ ] REMAINDER [ ]

$8 - 7 =$ [ ]

$6 \times 3 =$ [ ]

$8 + 8 =$ [ ]

$4 - 3 =$ [ ]

$9 \times 9 =$ [ ]

$8 - 3 =$ [ ]

$2 \times 7 =$ [ ]

$1 \div 1 =$ [ ]

$14 \div 2 =$ [ ]

$3 + 1 =$ [ ]

$7 \times 0 =$ [ ]

$9 - 4 =$ [ ]

$5 + 7 =$ [ ]

$2 \times 6 =$ [ ]

$4 + 2 =$ [ ]

$14 \div 7 =$ [ ]

$3 \times 6 =$ [ ]

$11 \div 3 =$ [ ] REMAINDER [ ]

$1 \times 9 =$ [ ]

$6 + 1 =$ [ ]

$8 - 5 =$ [ ]

$6 + 2 =$ [ ]

$12 \div 9 =$ [ ] REMAINDER [ ]

$16 - 8 =$ [ ]

$5 + 9 =$ [ ]

$18 - 9 =$ [ ]

$14 \div 3 =$ [ ] REMAINDER [ ]

$2 \times 2 =$ [ ]

$10 \div 2 =$ [ ]

$8 + 9 =$ [ ]

$2 \times 1 =$ [ ]

$19 \div 6 =$ [ ] REMAINDER [ ]

$14 - 7 =$ [ ]

$11 \div 8 =$ [ ] REMAINDER [ ]

$6 \times 4 =$ [ ]

$11 - 7 =$ [ ]

$5 \times 1 =$ [ ]

$12 - 3 =$ [ ]

$9 + 9 =$ [ ]

$14 - 8 =$ [ ]

$12 \div 4 =$ [ ]

$9 \times 8 =$ [ ]

$2 + 6 =$ [ ]

$0 + 4 =$ [ ]

$6 \div 1 =$ [ ]

$3 - 1 =$ [ ]

$4 + 6 =$ [ ]

$16 \div 8 =$ [ ]

$5 \times 6 =$ [ ]

End Time [ ] : [ ]

I. Counting Test

Measure the time required for you to count from 1 to 120 aloud as fast as you can.

☐ sec.

II. Word Memorization Test

Memorize as many words as you can **within two minutes.**

shadow	shape	opinion	history	baggage	barley
question	affair	shop	postcard	cherry	husband
lantern	change	longbow	seagull	scenery	pupil
floorboard	outset	medicine	spark	seaside	record
icicle	calendar	money	fraction	camellia	battle

Write out as many words as you can remember **in two minutes** on the back of this page. How many words can you remember?

Number of words memorized ☐ words

Word Memorization Test Answers

III. Stroop Test

Please take the **Stroop Test** for Week 5, located on page **viii** of the Appendix.

Start Time ☐ : ☐

$5 \times 9 =$ ☐

$1 + 2 =$ ☐

$16 \div 2 =$ ☐

$15 - 8 =$ ☐

$3 + 4 =$ ☐

$7 \times 9 =$ ☐

$12 - 9 =$ ☐

$6 \div 2 =$ ☐

$3 + 2 =$ ☐

$7 - 5 =$ ☐

$2 \times 9 =$ ☐

$16 \div 7 =$ ☐ REMAINDER ☐

$6 - 2 =$ ☐

$3 + 5 =$ ☐

$1 \times 1 =$ ☐

$9 \div 6 =$ ☐ REMAINDER ☐

$15 \div 5 =$ ☐

$0 + 8 =$ ☐

$8 \times 9 =$ ☐

$2 + 3 =$ ☐

$5 \times 7 =$ ☐

$8 \div 6 =$ ☐ REMAINDER ☐

$8 \times 5 =$ ☐

$11 - 4 =$ ☐

$15 - 9 =$ ☐

$11 \div 4 =$ ☐ REMAINDER ☐

$9 + 3 =$ ☐

$9 - 9 =$ ☐

$5 \times 4 =$ ☐

$8 + 6 =$ ☐

$6 \div 4 =$ ☐ REMAINDER ☐

$7 - 6 =$ ☐

$6 - 1 =$ ☐

$4 + 3 =$ ☐

$6 + 5 =$ ☐

$2 \times 4 =$ ☐

$16 \div 4 =$ ☐

$11 - 8 =$ ☐

$5 \times 5 =$ ☐

$8 - 6 =$ ☐

$1 \times 6 =$ ☐

$13 - 5 =$ ☐

$7 + 6 =$ ☐

$8 \div 4 =$ ☐

$6 + 7 =$ ☐

$18 \div 6 =$ ☐

$2 + 8 =$ ☐

$4 - 1 =$ ☐

$4 \times 9 =$ ☐

$15 \div 4 =$ ☐ REMAINDER ☐

$5 + 3 =$ ☐

$9 \div 8 =$ ☐ REMAINDER ☐

$6 \times 7 =$ ☐

$6 - 3 =$ ☐

$18 \div 2 =$ ☐

$7 + 4 =$ ☐

$5 - 1 =$ ☐

$6 \times 2 =$ ☐

$10 - 8 =$ ☐

$8 \times 1 =$ ☐

$4 \div 3 =$ ☐ REMAINDER ☐

$5 + 1 =$ ☐

$12 - 4 =$ ☐

$5 \times 8 =$ ☐

$2 + 5 =$ ☐

$9 \div 3 =$ ☐

$4 \times 0 =$ ☐

$17 - 9 =$ ☐

$1 \times 4 =$ ☐

$3 + 8 =$ ☐

$7 \div 5 =$ ☐ REMAINDER ☐

$9 - 5 =$ ☐

$7 + 9 =$ ☐

$9 \times 5 =$ ☐

$12 \div 2 =$ ☐

$8 \times 3 =$ ☐

$6 \div 6 =$ ☐

$8 + 7 =$ ☐

$16 - 9 =$ ☐

$2 + 4 =$ ☐

$4 \div 2 =$ ☐

$13 - 8 =$ ☐

$7 \times 8 =$ ☐

$18 \div 9 =$ ☐

$4 \times 2 =$ ☐

$10 \div 8 =$ ☐ REMAINDER ☐

$8 - 5 =$ ☐

$2 \times 3 =$ ☐

$8 \div 2 =$ ☐

$8 + 1 =$ ☐

$3 + 6 =$ ☐

$11 \div 5 =$ ☐ REMAINDER ☐

$4 \times 3 =$ ☐

$8 + 5 =$ ☐

$1 - 1 =$ ☐

$10 \div 3 =$ ☐ REMAINDER ☐

$8 \times 8 =$ ☐

$14 - 5 =$ ☐

$4 + 7 =$ ☐

$8 - 3 =$ ☐

End Time ☐ : ☐

74

Start Time ☐ : ☐

$18 \div 3 =$ ☐

$9 + 3 =$ ☐

$13 \div 8 =$ ☐ REMAINDER ☐

$9 \times 5 =$ ☐

$2 \times 3 =$ ☐

$2 + 5 =$ ☐

$7 + 8 =$ ☐

$7 \times 6 =$ ☐

$18 \div 9 =$ ☐

$11 - 8 =$ ☐

$12 - 7 =$ ☐

$1 \times 4 =$ ☐

$3 + 6 =$ ☐

$14 - 7 =$ ☐

$16 \div 4 =$ ☐

$7 - 3 =$ ☐

$8 \times 6 =$ ☐

$10 - 7 =$ ☐

$9 \div 5 =$ ☐ REMAINDER ☐

$6 \times 9 =$ ☐

$7 - 4 =$ ☐

$3 + 1 =$ ☐

$6 + 8 =$ ☐

$7 \times 5 =$ ☐

$2 - 2 =$ ☐

$13 \div 5 =$ ☐ REMAINDER ☐

$5 + 3 =$ ☐

$5 \times 5 =$ ☐

$6 - 6 =$ ☐

$9 \div 1 =$ ☐

$3 + 0 =$ ☐

$17 \div 7 =$ ☐ REMAINDER ☐

$7 + 4 =$ ☐

$11 \div 7 =$ ☐ REMAINDER ☐

$2 + 7 =$ ☐

$13 - 4 =$ ☐

$19 \div 9 =$ ☐ REMAINDER ☐

$9 \times 2 =$ ☐

$5 + 7 =$ ☐

$12 \div 6 =$ ☐

$2 \times 4 =$ ☐

$13 - 9 =$ ☐

$7 - 1 =$ ☐

$5 \times 2 =$ ☐

$8 + 2 =$ ☐

$9 - 5 =$ ☐

$8 \times 3 =$ ☐

$8 - 8 =$ ☐

$12 \div 4 =$ ☐

$5 + 1 =$ ☐

$2 \times 8 =$ [ ]

$8 \div 2 =$ [ ]

$8 + 7 =$ [ ]

$4 + 2 =$ [ ]

$11 - 9 =$ [ ]

$12 \div 2 =$ [ ]

$5 \times 9 =$ [ ]

$4 - 1 =$ [ ]

$14 \div 2 =$ [ ]

$4 + 8 =$ [ ]

$5 \div 1 =$ [ ]

$2 \times 7 =$ [ ]

$9 - 8 =$ [ ]

$2 + 1 =$ [ ]

$1 \times 9 =$ [ ]

$6 \div 5 =$ [ ] REMAINDER [ ]

$14 - 8 =$ [ ]

$14 \div 9 =$ [ ] REMAINDER [ ]

$6 + 3 =$ [ ]

$17 - 9 =$ [ ]

$4 + 1 =$ [ ]

$4 \times 7 =$ [ ]

$5 \div 3 =$ [ ] REMAINDER [ ]

$16 - 9 =$ [ ]

$9 \times 3 =$ [ ]

$3 + 7 =$ [ ]

$12 \div 3 =$ [ ]

$0 \times 5 =$ [ ]

$6 \times 8 =$ [ ]

$6 - 2 =$ [ ]

$5 + 2 =$ [ ]

$16 \div 3 =$ [ ] REMAINDER [ ]

$11 - 5 =$ [ ]

$3 \times 8 =$ [ ]

$9 \div 3 =$ [ ]

$7 - 7 =$ [ ]

$13 \div 2 =$ [ ] REMAINDER [ ]

$1 + 7 =$ [ ]

$9 \times 7 =$ [ ]

$8 - 5 =$ [ ]

$5 \div 5 =$ [ ]

$3 + 8 =$ [ ]

$4 \times 8 =$ [ ]

$5 \div 4 =$ [ ] REMAINDER [ ]

$5 + 9 =$ [ ]

$8 - 2 =$ [ ]

$8 + 8 =$ [ ]

$8 \times 7 =$ [ ]

$4 \times 6 =$ [ ]

$16 - 7 =$ [ ]

End Time [ ] : [ ]

Start Time ☐ : ☐

10 − 8 = ☐

15 ÷ 3 = ☐

9 + 0 = ☐

15 ÷ 9 = ☐ REMAINDER ☐

9 − 1 = ☐

2 × 2 = ☐

2 + 3 = ☐

8 × 7 = ☐

11 ÷ 5 = ☐ REMAINDER ☐

4 ÷ 2 = ☐

5 + 4 = ☐

10 − 1 = ☐

6 × 2 = ☐

5 − 1 = ☐

2 × 5 = ☐

8 + 5 = ☐

7 + 9 = ☐

17 − 8 = ☐

10 ÷ 5 = ☐

4 − 4 = ☐

19 ÷ 2 = ☐ REMAINDER ☐

3 × 6 = ☐

1 + 5 = ☐

5 ÷ 2 = ☐ REMAINDER ☐

4 + 7 = ☐

12 − 8 = ☐

1 × 3 = ☐

3 × 7 = ☐

6 + 6 = ☐

5 × 4 = ☐

14 ÷ 8 = ☐ REMAINDER ☐

7 + 2 = ☐

3 × 1 = ☐

10 ÷ 2 = ☐

3 + 3 = ☐

12 − 5 = ☐

10 − 3 = ☐

18 ÷ 2 = ☐

8 − 4 = ☐

3 + 5 = ☐

8 × 1 = ☐

9 + 2 = ☐

6 × 1 = ☐

2 − 0 = ☐

6 + 7 = ☐

3 ÷ 3 = ☐

6 − 1 = ☐

8 × 4 = ☐

6 − 4 = ☐

13 ÷ 3 = ☐ REMAINDER ☐

$3 \times 3 =$ ☐

$1 + 4 =$ ☐

$17 \div 8 =$ ☐ REMAINDER ☐

$7 - 7 =$ ☐

$3 \times 5 =$ ☐

$17 \div 2 =$ ☐ REMAINDER ☐

$9 - 2 =$ ☐

$6 + 1 =$ ☐

$14 - 9 =$ ☐

$3 + 4 =$ ☐

$8 \times 9 =$ ☐

$1 \times 8 =$ ☐

$9 + 4 =$ ☐

$8 \div 4 =$ ☐

$12 - 3 =$ ☐

$10 \div 8 =$ ☐ REMAINDER ☐

$7 \times 9 =$ ☐

$4 \times 9 =$ ☐

$15 \div 5 =$ ☐

$5 - 5 =$ ☐

$4 + 3 =$ ☐

$15 \div 7 =$ ☐ REMAINDER ☐

$5 \times 3 =$ ☐

$8 + 9 =$ ☐

$4 \div 4 =$ ☐

$7 \times 0 =$ ☐

$8 - 2 =$ ☐

$8 + 6 =$ ☐

$13 - 6 =$ ☐

$6 \div 2 =$ ☐

$8 + 3 =$ ☐

$4 \times 4 =$ ☐

$16 \div 2 =$ ☐

$9 - 3 =$ ☐

$18 - 9 =$ ☐

$2 + 4 =$ ☐

$6 \times 4 =$ ☐

$7 + 3 =$ ☐

$15 \div 6 =$ ☐ REMAINDER ☐

$8 + 1 =$ ☐

$11 - 3 =$ ☐

$16 \div 8 =$ ☐

$12 - 9 =$ ☐

$9 \times 9 =$ ☐

$7 \times 1 =$ ☐

$3 + 9 =$ ☐

$6 \div 3 =$ ☐

$8 \div 6 =$ ☐ REMAINDER ☐

$2 \times 9 =$ ☐

$1 - 1 =$ ☐

End Time ☐ : ☐

Start Time ☐ : ☐

$8 \div 3 =$ ☐ REMAINDER ☐

$5 \times 9 =$ ☐

$6 + 1 =$ ☐

$3 - 2 =$ ☐

$17 \div 3 =$ ☐ REMAINDER ☐

$14 - 5 =$ ☐

$4 + 5 =$ ☐

$2 \div 2 =$ ☐

$3 \times 3 =$ ☐

$2 + 0 =$ ☐

$7 - 1 =$ ☐

$1 \times 9 =$ ☐

$6 \div 2 =$ ☐

$5 + 7 =$ ☐

$5 \times 4 =$ ☐

$6 - 3 =$ ☐

$5 + 6 =$ ☐

$5 \times 3 =$ ☐

$13 \div 8 =$ ☐ REMAINDER ☐

$8 - 7 =$ ☐

$2 + 6 =$ ☐

$10 - 3 =$ ☐

$8 \times 4 =$ ☐

$4 + 3 =$ ☐

$7 + 7 =$ ☐

$9 \div 8 =$ ☐ REMAINDER ☐

$0 \times 4 =$ ☐

$4 \div 2 =$ ☐

$11 - 4 =$ ☐

$7 - 3 =$ ☐

$5 + 1 =$ ☐

$8 \div 2 =$ ☐

$5 \times 7 =$ ☐

$11 - 9 =$ ☐

$9 \times 7 =$ ☐

$8 \div 7 =$ ☐ REMAINDER ☐

$6 \times 4 =$ ☐

$8 - 5 =$ ☐

$4 + 7 =$ ☐

$5 \div 1 =$ ☐

$10 \div 5 =$ ☐

$9 + 9 =$ ☐

$8 \times 6 =$ ☐

$2 + 4 =$ ☐

$10 - 4 =$ ☐

$6 \times 5 =$ ☐

$15 \div 6 =$ ☐ REMAINDER ☐

$5 - 2 =$ ☐

$11 - 2 =$ ☐

$8 + 5 =$ ☐

$11 \div 9 =$ ☐ REMAINDER ☐　　$6 - 5 =$ ☐　　$8 + 1 =$ ☐

$0 \times 9 =$ ☐　　$2 + 2 =$ ☐　　$13 \div 7 =$ ☐ REMAINDER ☐

$9 - 6 =$ ☐　　$10 \div 2 =$ ☐　　$2 - 1 =$ ☐

$1 \times 7 =$ ☐　　$7 \times 9 =$ ☐　　$2 \times 7 =$ ☐

$5 + 9 =$ ☐　　$7 + 6 =$ ☐　　$9 \div 3 =$ ☐

$16 \div 8 =$ ☐　　$4 \times 3 =$ ☐　　$12 - 7 =$ ☐

$10 - 1 =$ ☐　　$10 - 8 =$ ☐　　$2 \times 4 =$ ☐

$17 - 8 =$ ☐　　$5 \times 8 =$ ☐　　$1 + 1 =$ ☐

$3 \times 6 =$ ☐　　$14 \div 2 =$ ☐　　$5 - 3 =$ ☐

$1 + 7 =$ ☐　　$5 + 3 =$ ☐　　$3 \times 2 =$ ☐

$13 \div 5 =$ ☐ REMAINDER ☐　　$15 - 8 =$ ☐　　$13 \div 9 =$ ☐ REMAINDER ☐

$11 - 3 =$ ☐　　$7 \times 6 =$ ☐　　$6 + 7 =$ ☐

$9 + 1 =$ ☐　　$11 \div 6 =$ ☐ REMAINDER ☐　　$7 \div 5 =$ ☐ REMAINDER ☐

$15 \div 3 =$ ☐　　$7 - 4 =$ ☐　　$6 \times 2 =$ ☐

$6 \times 6 =$ ☐　　$0 + 4 =$ ☐　　$6 + 5 =$ ☐

$9 + 3 =$ ☐　　$8 \times 7 =$ ☐　　$5 - 1 =$ ☐

$12 \div 3 =$ ☐　　$3 \div 1 =$ ☐　　**End Time** ☐ : ☐

Start Time ☐ : ☐

$3 + 1 =$ ☐

$8 \times 5 =$ ☐

$9 \div 6 =$ ☐ REMAINDER ☐

$8 - 4 =$ ☐

$7 \times 2 =$ ☐

$6 \div 6 =$ ☐

$12 - 8 =$ ☐

$3 \times 7 =$ ☐

$11 \div 7 =$ ☐ REMAINDER ☐

$3 + 3 =$ ☐

$19 \div 6 =$ ☐ REMAINDER ☐

$5 - 5 =$ ☐

$4 + 6 =$ ☐

$6 + 8 =$ ☐

$1 - 1 =$ ☐

$4 \times 6 =$ ☐

$7 \times 1 =$ ☐

$13 - 8 =$ ☐

$1 + 2 =$ ☐

$6 \times 1 =$ ☐

$18 \div 7 =$ ☐ REMAINDER ☐

$4 - 1 =$ ☐

$7 + 5 =$ ☐

$4 \times 1 =$ ☐

$16 - 8 =$ ☐

$11 \div 4 =$ ☐ REMAINDER ☐

$3 + 4 =$ ☐

$4 - 2 =$ ☐

$12 - 5 =$ ☐

$14 \div 7 =$ ☐

$9 \times 5 =$ ☐

$3 + 5 =$ ☐

$16 \div 4 =$ ☐

$6 - 1 =$ ☐

$2 \times 3 =$ ☐

$9 \times 8 =$ ☐

$1 + 9 =$ ☐

$18 \div 6 =$ ☐

$2 \times 9 =$ ☐

$8 + 4 =$ ☐

$17 \div 4 =$ ☐ REMAINDER ☐

$13 - 5 =$ ☐

$4 + 4 =$ ☐

$16 \div 2 =$ ☐

$18 - 9 =$ ☐

$2 + 1 =$ ☐

$9 + 7 =$ ☐

$7 \times 3 =$ ☐

$4 - 3 =$ ☐

$18 \div 3 =$ ☐

$8 - 8 =$ ☐

$15 \div 3 =$ ☐

$2 + 8 =$ ☐

$2 \times 6 =$ ☐

$6 - 4 =$ ☐

$8 \times 1 =$ ☐

$4 + 2 =$ ☐

$9 - 8 =$ ☐

$2 \times 5 =$ ☐

$7 + 9 =$ ☐

$4 \div 3 =$ ☐ REMAINDER ☐

$5 \times 5 =$ ☐

$2 + 9 =$ ☐

$12 \div 2 =$ ☐

$14 - 8 =$ ☐

$4 \times 8 =$ ☐

$6 \div 1 =$ ☐

$7 + 2 =$ ☐

$19 \div 2 =$ ☐ REMAINDER ☐

$13 - 9 =$ ☐

$5 + 4 =$ ☐

$1 \times 5 =$ ☐

$8 \div 8 =$ ☐

$5 \times 1 =$ ☐

$10 - 2 =$ ☐

$17 \div 6 =$ ☐ REMAINDER ☐

$7 + 0 =$ ☐

$2 \times 2 =$ ☐

$5 - 1 =$ ☐

$6 + 4 =$ ☐

$17 \div 8 =$ ☐ REMAINDER ☐

$12 - 6 =$ ☐

$2 \times 0 =$ ☐

$18 \div 9 =$ ☐

$12 - 9 =$ ☐

$8 + 8 =$ ☐

$8 \div 4 =$ ☐

$9 \times 9 =$ ☐

$15 - 6 =$ ☐

$5 \times 6 =$ ☐

$18 \div 2 =$ ☐

$2 - 2 =$ ☐

$6 \times 8 =$ ☐

$6 + 3 =$ ☐

$10 \div 3 =$ ☐ REMAINDER ☐

$6 \times 3 =$ ☐

$2 + 5 =$ ☐

$7 - 4 =$ ☐

$6 + 9 =$ ☐

$3 \div 2 =$ ☐ REMAINDER ☐

End Time ☐ : ☐

I. Counting Test

Measure the time required for you to count from 1 to 120 aloud as fast as you can.

☐ sec.

II. Word Memorization Test

Memorize as many words as you can **within two minutes.**

bathroom	left	battery	contest	handbag	clumsy
cockerel	hearth	grammar	hat	eggcup	ancestor
rabbit	performance	cavity	watermelon	break	snooze
bubble	Orient	glasses	fireworks	emotion	lotus
strength	ring	heat	wallet	sunrise	response

Write out as many words as you can remember **in two minutes** on the back of this page. How many words can you remember?

Number of words memorized ☐ words

Word Memorization Test Answers

III. Stroop Test

Please take the **Stroop Test** for Week 6, located on page **ix** of the Appendix.

Start Time ☐ : ☐

$3 + 4 =$ ☐

$7 \div 1 =$ ☐

$7 - 5 =$ ☐

$10 \div 7 =$ ☐ REMAINDER ☐

$5 + 1 =$ ☐

$15 - 8 =$ ☐

$1 \times 3 =$ ☐

$10 - 3 =$ ☐

$4 \times 8 =$ ☐

$17 \div 3 =$ ☐ REMAINDER ☐

$5 \times 7 =$ ☐

$6 + 3 =$ ☐

$10 - 5 =$ ☐

$14 \div 2 =$ ☐

$7 + 8 =$ ☐

$5 \times 6 =$ ☐

$19 \div 3 =$ ☐ REMAINDER ☐

$9 + 5 =$ ☐

$1 \times 5 =$ ☐

$11 - 9 =$ ☐

$12 \div 2 =$ ☐

$5 \times 8 =$ ☐

$5 + 4 =$ ☐

$12 - 8 =$ ☐

$7 \div 3 =$ ☐ REMAINDER ☐

$5 \times 1 =$ ☐

$3 - 3 =$ ☐

$2 + 9 =$ ☐

$12 \div 4 =$ ☐

$7 - 1 =$ ☐

$3 \times 1 =$ ☐

$4 + 5 =$ ☐

$14 - 6 =$ ☐

$2 + 7 =$ ☐

$6 - 5 =$ ☐

$8 \times 5 =$ ☐

$12 \div 7 =$ ☐ REMAINDER ☐

$2 \times 8 =$ ☐

$7 - 2 =$ ☐

$5 + 8 =$ ☐

$8 \div 4 =$ ☐

$5 - 1 =$ ☐

$2 \times 2 =$ ☐

$18 \div 2 =$ ☐

$9 + 4 =$ ☐

$2 + 2 =$ ☐

$9 - 4 =$ ☐

$0 \times 9 =$ ☐

$15 \div 8 =$ ☐ REMAINDER ☐

$4 + 8 =$ ☐

$3 \times 4 =$ ☐

$16 - 7 =$ ☐

$9 + 2 =$ ☐

$13 - 9 =$ ☐

$12 \div 3 =$ ☐

$1 \times 9 =$ ☐

$7 + 2 =$ ☐

$15 \div 2 =$ ☐ REMAINDER ☐

$9 - 6 =$ ☐

$4 \times 3 =$ ☐

$8 + 3 =$ ☐

$12 \div 6 =$ ☐

$7 + 9 =$ ☐

$8 \times 6 =$ ☐

$13 - 6 =$ ☐

$5 \div 1 =$ ☐

$1 + 8 =$ ☐

$18 - 9 =$ ☐

$5 + 9 =$ ☐

$11 \div 3 =$ ☐ REMAINDER ☐

$4 \times 7 =$ ☐

$6 \div 3 =$ ☐

$15 - 6 =$ ☐

$2 \times 1 =$ ☐

$11 \div 2 =$ ☐ REMAINDER ☐

$5 \times 9 =$ ☐

$3 + 2 =$ ☐

$18 \div 3 =$ ☐

$5 - 3 =$ ☐

$8 - 4 =$ ☐

$5 + 7 =$ ☐

$19 \div 8 =$ ☐ REMAINDER ☐

$6 \times 3 =$ ☐

$2 + 6 =$ ☐

$19 \div 9 =$ ☐ REMAINDER ☐

$9 \times 2 =$ ☐

$6 - 4 =$ ☐

$5 \times 3 =$ ☐

$9 + 7 =$ ☐

$14 - 8 =$ ☐

$9 \div 9 =$ ☐

$7 - 3 =$ ☐

$9 \times 6 =$ ☐

$4 \times 1 =$ ☐

$5 \div 3 =$ ☐ REMAINDER ☐

$2 - 1 =$ ☐

$0 + 8 =$ ☐

$3 \times 5 =$ ☐

$8 \div 2 =$ ☐

$4 + 3 =$ ☐

End Time ☐ : ☐

Start Time ☐ : ☐

$2 + 5 =$ ☐

$8 - 3 =$ ☐

$5 \div 5 =$ ☐

$14 - 5 =$ ☐

$4 \times 2 =$ ☐

$12 \div 2 =$ ☐

$3 \times 9 =$ ☐

$15 - 9 =$ ☐

$7 \times 4 =$ ☐

$6 + 5 =$ ☐

$18 \div 6 =$ ☐

$11 \div 5 =$ ☐ REMAINDER ☐

$8 + 1 =$ ☐

$9 - 5 =$ ☐

$2 + 3 =$ ☐

$2 \times 6 =$ ☐

$5 - 2 =$ ☐

$16 \div 5 =$ ☐ REMAINDER ☐

$2 - 2 =$ ☐

$1 + 5 =$ ☐

$7 \times 2 =$ ☐

$19 \div 7 =$ ☐ REMAINDER ☐

$14 - 7 =$ ☐

$6 \times 8 =$ ☐

$3 + 1 =$ ☐

$9 \div 3 =$ ☐

$1 \times 2 =$ ☐

$19 \div 4 =$ ☐ REMAINDER ☐

$8 + 6 =$ ☐

$11 - 4 =$ ☐

$4 \times 6 =$ ☐

$4 + 7 =$ ☐

$13 - 7 =$ ☐

$4 \div 2 =$ ☐

$5 - 5 =$ ☐

$1 + 2 =$ ☐

$18 \div 9 =$ ☐

$2 \times 4 =$ ☐

$3 + 3 =$ ☐

$7 + 6 =$ ☐

$13 - 8 =$ ☐

$17 \div 6 =$ ☐ REMAINDER ☐

$9 + 8 =$ ☐

$5 \times 2 =$ ☐

$2 \times 3 =$ ☐

$9 \div 7 =$ ☐ REMAINDER ☐

$7 + 7 =$ ☐

$9 \times 5 =$ ☐

$4 - 1 =$ ☐

$9 - 0 =$ ☐

Time Required ☐ : ☐

13 − 5 = ☐

7 + 0 = ☐

9 × 3 = ☐

10 − 7 = ☐

16 ÷ 7 = ☐ REMAINDER ☐

2 × 9 = ☐

14 − 9 = ☐

4 + 2 = ☐

8 × 3 = ☐

11 ÷ 6 = ☐ REMAINDER ☐

18 ÷ 7 = ☐ REMAINDER ☐

8 + 7 = ☐

9 × 7 = ☐

16 − 9 = ☐

10 ÷ 5 = ☐

1 + 6 = ☐

8 × 8 = ☐

4 + 6 = ☐

7 ÷ 7 = ☐

8 − 1 = ☐

16 ÷ 8 = ☐

2 + 4 = ☐

8 − 5 = ☐

0 × 6 = ☐

6 − 1 = ☐

5 + 5 = ☐

7 × 3 = ☐

1 ÷ 1 = ☐

6 ÷ 2 = ☐

9 × 8 = ☐

12 − 4 = ☐

2 + 7 = ☐

4 × 5 = ☐

8 ÷ 3 = ☐ REMAINDER ☐

2 − 1 = ☐

6 × 9 = ☐

8 + 5 = ☐

3 × 6 = ☐

10 ÷ 2 = ☐

6 − 4 = ☐

1 + 3 = ☐

1 × 6 = ☐

8 ÷ 5 = ☐ REMAINDER ☐

6 + 6 = ☐

16 ÷ 2 = ☐

9 − 1 = ☐

2 × 7 = ☐

10 − 6 = ☐

7 + 5 = ☐

13 ÷ 3 = ☐ REMAINDER ☐

End Time ☐ : ☐

Start Time ☐ : ☐

14 − 9 = ☐

4 ÷ 4 = ☐

6 × 2 = ☐

12 − 7 = ☐

3 + 8 = ☐

5 × 7 = ☐

6 + 7 = ☐

7 ÷ 6 = ☐ REMAINDER ☐

6 − 2 = ☐

6 + 5 = ☐

17 ÷ 5 = ☐ REMAINDER ☐

6 × 4 = ☐

6 + 3 = ☐

12 − 6 = ☐

3 ÷ 2 = ☐ REMAINDER ☐

6 × 3 = ☐

8 ÷ 2 = ☐

2 + 3 = ☐

2 × 8 = ☐

12 ÷ 2 = ☐

4 + 7 = ☐

14 − 6 = ☐

9 − 1 = ☐

8 × 9 = ☐

2 + 7 = ☐

12 ÷ 3 = ☐

3 × 2 = ☐

4 − 0 = ☐

1 × 4 = ☐

9 − 6 = ☐

4 + 9 = ☐

8 ÷ 6 = ☐ REMAINDER ☐

4 − 4 = ☐

6 + 2 = ☐

2 × 1 = ☐

1 + 5 = ☐

14 ÷ 7 = ☐

8 − 6 = ☐

3 × 4 = ☐

16 ÷ 3 = ☐ REMAINDER ☐

13 − 8 = ☐

5 − 1 = ☐

8 + 5 = ☐

11 ÷ 8 = ☐ REMAINDER ☐

2 + 5 = ☐

4 × 5 = ☐

11 − 4 = ☐

9 × 4 = ☐

8 ÷ 8 = ☐

2 + 6 = ☐

$11 \div 6 =$ ☐ REMAINDER ☐ $14 \div 4 =$ ☐ REMAINDER ☐ $4 + 2 =$ ☐

$3 - 2 =$ ☐ $2 \times 5 =$ ☐ $12 \div 5 =$ ☐ REMAINDER ☐

$5 + 6 =$ ☐ $18 \div 3 =$ ☐ $11 - 9 =$ ☐

$4 + 3 =$ ☐ $2 + 9 =$ ☐ $7 \times 2 =$ ☐

$4 \times 6 =$ ☐ $4 \times 3 =$ ☐ $0 + 5 =$ ☐

$12 \div 9 =$ ☐ REMAINDER ☐ $10 - 6 =$ ☐ $15 \div 3 =$ ☐

$5 \times 5 =$ ☐ $1 + 9 =$ ☐ $9 + 4 =$ ☐

$10 \div 2 =$ ☐ $2 \times 6 =$ ☐ $3 \times 8 =$ ☐

$3 + 2 =$ ☐ $4 - 2 =$ ☐ $15 \div 5 =$ ☐

$7 \times 7 =$ ☐ $6 \div 1 =$ ☐ $7 - 3 =$ ☐

$13 - 7 =$ ☐ $2 + 2 =$ ☐ $6 \times 9 =$ ☐

$14 \div 2 =$ ☐ $9 \times 8 =$ ☐ $16 \div 7 =$ ☐ REMAINDER ☐

$9 + 5 =$ ☐ $6 - 5 =$ ☐ $14 - 8 =$ ☐

$12 \div 6 =$ ☐ $5 \times 1 =$ ☐ $7 \times 3 =$ ☐

$10 - 2 =$ ☐ $17 \div 8 =$ ☐ REMAINDER ☐ $3 + 6 =$ ☐

$9 \times 2 =$ ☐ $8 - 5 =$ ☐ $1 - 0 =$ ☐

$13 - 9 =$ ☐ $9 + 2 =$ ☐   **End Time** ☐ : ☐

Start Time ☐ : ☐

$1 + 2 =$ ☐

$9 - 8 =$ ☐

$14 \div 9 =$ ☐ REMAINDER ☐

$6 \times 5 =$ ☐

$14 \div 6 =$ ☐ REMAINDER ☐

$8 + 6 =$ ☐

$4 \times 9 =$ ☐

$18 - 9 =$ ☐

$11 - 8 =$ ☐

$18 \div 2 =$ ☐

$4 + 1 =$ ☐

$8 \times 8 =$ ☐

$12 \div 2 =$ ☐

$9 \times 7 =$ ☐

$7 + 7 =$ ☐

$2 - 1 =$ ☐

$12 \div 8 =$ ☐ REMAINDER ☐

$3 \times 9 =$ ☐

$5 + 1 =$ ☐

$12 - 4 =$ ☐

$5 + 8 =$ ☐

$8 - 8 =$ ☐

$18 \div 4 =$ ☐ REMAINDER ☐

$1 \times 5 =$ ☐

$7 \times 9 =$ ☐

$5 + 4 =$ ☐

$9 - 4 =$ ☐

$18 \div 6 =$ ☐

$12 - 8 =$ ☐

$5 \times 7 =$ ☐

$9 + 1 =$ ☐

$16 \div 2 =$ ☐

$9 + 9 =$ ☐

$4 \div 3 =$ ☐ REMAINDER ☐

$11 - 3 =$ ☐

$0 + 7 =$ ☐

$3 \times 5 =$ ☐

$7 + 4 =$ ☐

$11 \div 7 =$ ☐ REMAINDER ☐

$12 - 3 =$ ☐

$7 \times 6 =$ ☐

$8 - 4 =$ ☐

$3 \times 6 =$ ☐

$9 - 9 =$ ☐

$7 \div 7 =$ ☐

$3 + 5 =$ ☐

$0 + 1 =$ ☐

$9 \times 5 =$ ☐

$6 - 3 =$ ☐

$10 \div 5 =$ ☐

$4 \times 7 =$ ☐

$4 \div 1 =$ ☐

$1 + 3 =$ ☐

$10 - 4 =$ ☐

$15 - 6 =$ ☐

$4 \times 4 =$ ☐

$14 - 5 =$ ☐

$13 \div 4 =$ ☐ REMAINDER ☐

$6 + 6 =$ ☐

$15 - 9 =$ ☐

$18 \div 7 =$ ☐ REMAINDER ☐

$3 + 4 =$ ☐

$16 \div 4 =$ ☐

$4 \times 2 =$ ☐

$1 + 6 =$ ☐

$7 - 4 =$ ☐

$5 \times 8 =$ ☐

$1 + 1 =$ ☐

$13 \div 6 =$ ☐ REMAINDER ☐

$1 \times 6 =$ ☐

$6 \div 6 =$ ☐

$9 - 7 =$ ☐

$8 + 3 =$ ☐

$2 \div 2 =$ ☐

$7 \times 4 =$ ☐

$13 - 5 =$ ☐

$2 + 4 =$ ☐

$6 \times 1 =$ ☐

$8 - 3 =$ ☐

$2 \times 3 =$ ☐

$5 - 3 =$ ☐

$9 \div 3 =$ ☐

$8 + 9 =$ ☐

$6 \times 6 =$ ☐

$4 + 6 =$ ☐

$17 \div 7 =$ ☐ REMAINDER ☐

$4 + 5 =$ ☐

$8 \times 6 =$ ☐

$15 \div 8 =$ ☐ REMAINDER ☐

$8 + 8 =$ ☐

$14 \div 3 =$ ☐ REMAINDER ☐

$12 \div 4 =$ ☐

$6 \times 7 =$ ☐

$4 - 3 =$ ☐

$3 \times 0 =$ ☐

$6 + 9 =$ ☐

$16 - 7 =$ ☐

$16 \div 8 =$ ☐

$8 \times 4 =$ ☐

$7 - 0 =$ ☐

End Time ☐ : ☐

Day 35

Start Time ☐ : ☐

$4 - 1 =$ ☐

$1 \times 2 =$ ☐

$3 + 2 =$ ☐

$5 \div 3 =$ ☐ REMAINDER ☐

$8 + 2 =$ ☐

$6 \div 1 =$ ☐

$7 \times 3 =$ ☐

$5 + 6 =$ ☐

$9 - 2 =$ ☐

$3 \times 8 =$ ☐

$9 + 2 =$ ☐

$18 \div 9 =$ ☐

$9 - 7 =$ ☐

$3 \times 7 =$ ☐

$9 \div 4 =$ ☐ REMAINDER ☐

$8 - 1 =$ ☐

$9 + 5 =$ ☐

$16 - 8 =$ ☐

$9 \times 0 =$ ☐

$4 + 2 =$ ☐

$7 \div 6 =$ ☐ REMAINDER ☐

$5 - 1 =$ ☐

$9 \times 5 =$ ☐

$8 \div 4 =$ ☐

$0 + 4 =$ ☐

$11 - 6 =$ ☐

$6 \times 2 =$ ☐

$10 \div 3 =$ ☐ REMAINDER ☐

$6 + 1 =$ ☐

$3 + 5 =$ ☐

$15 - 9 =$ ☐

$8 \div 2 =$ ☐

$7 \times 2 =$ ☐

$4 + 9 =$ ☐

$18 \div 5 =$ ☐ REMAINDER ☐

$8 \times 9 =$ ☐

$6 + 2 =$ ☐

$7 \times 6 =$ ☐

$9 - 8 =$ ☐

$4 \div 4 =$ ☐

$9 - 6 =$ ☐

$1 + 8 =$ ☐

$8 + 5 =$ ☐

$17 - 8 =$ ☐

$5 \times 3 =$ ☐

$12 - 6 =$ ☐

$10 - 8 =$ ☐

$15 \div 5 =$ ☐

$4 \times 9 =$ ☐

$16 \div 6 =$ ☐ REMAINDER ☐

$12 \div 6 =$ ☐

$14 - 6 =$ ☐

$3 + 6 =$ ☐

$7 \times 7 =$ ☐

$10 \div 8 =$ ☐ REMAINDER ☐

$6 \times 5 =$ ☐

$15 - 6 =$ ☐

$5 + 4 =$ ☐

$6 \div 4 =$ ☐ REMAINDER ☐

$2 \times 9 =$ ☐

$8 - 3 =$ ☐

$1 + 4 =$ ☐

$3 - 2 =$ ☐

$1 \times 9 =$ ☐

$12 \div 2 =$ ☐

$7 + 7 =$ ☐

$9 \div 7 =$ ☐ REMAINDER ☐

$2 - 1 =$ ☐

$3 \times 9 =$ ☐

$16 \div 4 =$ ☐

$6 - 2 =$ ☐

$7 + 6 =$ ☐

$9 \times 9 =$ ☐

$2 \times 6 =$ ☐

$13 - 8 =$ ☐

$12 \div 3 =$ ☐

$5 + 3 =$ ☐

$2 \times 8 =$ ☐

$13 \div 9 =$ ☐ REMAINDER ☐

$14 - 9 =$ ☐

$6 + 3 =$ ☐

$5 \times 4 =$ ☐

$5 + 8 =$ ☐

$12 \div 8 =$ ☐ REMAINDER ☐

$18 \div 3 =$ ☐

$9 \times 2 =$ ☐

$5 - 3 =$ ☐

$12 \div 4 =$ ☐

$9 + 6 =$ ☐

$6 + 7 =$ ☐

$13 - 5 =$ ☐

$2 + 7 =$ ☐

$16 \div 7 =$ ☐ REMAINDER ☐

$5 \times 2 =$ ☐

$5 - 2 =$ ☐

$1 \times 8 =$ ☐

$13 - 4 =$ ☐

$6 + 6 =$ ☐

$7 \div 7 =$ ☐

$7 \times 9 =$ ☐

End Time ☐ : ☐

I. Counting Test

Measure the time required for you to count from 1 to 120 aloud as fast as you can.

☐ sec.

II. Word Memorization Test

Memorize as many words as you can **within two minutes**.

wing	anniversary	season	centipede	whale	radish
smile	suitcase	yesterday	individual	mirror	festival
tangerine	pasture	strawberry	antenna	dance	game
container	dusk	dumpling	vicinity	request	sunshine
noodles	play	barefoot	telephone	sunset	dragonfly

Write out as many words as you can remember **in two minutes** on the back of this page. How many words can you remember?

Number of words memorized ☐ words

Word Memorization Test Answers

III. Stroop Test

Please take the **Stroop Test** for Week 7, located on page **x** of the Appendix.

Date ☐ M ☐ D

Start Time ☐ : ☐

$2 \times 2 =$ ☐

$18 \div 6 =$ ☐

$7 \times 4 =$ ☐

$10 - 9 =$ ☐

$2 + 3 =$ ☐

$10 - 4 =$ ☐

$7 \div 5 =$ ☐ REMAINDER ☐

$8 + 9 =$ ☐

$9 \div 3 =$ ☐

$7 + 3 =$ ☐

$7 - 2 =$ ☐

$9 \times 7 =$ ☐

$9 \div 5 =$ ☐ REMAINDER ☐

$1 + 6 =$ ☐

$6 \times 9 =$ ☐

$6 - 1 =$ ☐

$4 - 2 =$ ☐

$17 \div 6 =$ ☐ REMAINDER ☐

$11 - 4 =$ ☐

$6 \times 7 =$ ☐

$9 + 0 =$ ☐

$3 - 1 =$ ☐

$9 + 3 =$ ☐

$6 + 8 =$ ☐

$6 \div 6 =$ ☐

$9 + 8 =$ ☐

$3 \times 4 =$ ☐

$18 \div 2 =$ ☐

$7 - 3 =$ ☐

$6 \times 4 =$ ☐

$3 \div 3 =$ ☐

$8 \times 7 =$ ☐

$2 + 2 =$ ☐

$10 - 7 =$ ☐

$4 + 5 =$ ☐

$3 \times 2 =$ ☐

$12 - 9 =$ ☐

$7 \div 3 =$ ☐ REMAINDER ☐

$8 - 7 =$ ☐

$3 + 4 =$ ☐

$4 \times 4 =$ ☐

$10 \div 5 =$ ☐

$3 + 6 =$ ☐

$16 - 9 =$ ☐

$12 \div 7 =$ ☐ REMAINDER ☐

$5 \times 7 =$ ☐

$3 \times 3 =$ ☐

$19 \div 7 =$ ☐ REMAINDER ☐

$8 + 4 =$ ☐

$7 - 1 =$ ☐

97

$19 \div 4 =$ [ ] REMAINDER [ ]

$1 + 5 =$ [ ]

$4 \div 2 =$ [ ]

$5 - 3 =$ [ ]

$15 - 8 =$ [ ]

$4 + 8 =$ [ ]

$5 + 7 =$ [ ]

$4 \div 3 =$ [ ] REMAINDER [ ]

$5 - 4 =$ [ ]

$8 \times 3 =$ [ ]

$6 \times 8 =$ [ ]

$6 \div 2 =$ [ ]

$2 + 4 =$ [ ]

$6 \div 3 =$ [ ]

$5 + 5 =$ [ ]

$11 - 9 =$ [ ]

$8 \times 5 =$ [ ]

$9 \times 8 =$ [ ]

$8 \times 8 =$ [ ]

$9 \div 9 =$ [ ]

$6 - 5 =$ [ ]

$17 \div 7 =$ [ ] REMAINDER [ ]

$1 + 7 =$ [ ]

$4 \times 2 =$ [ ]

$10 - 6 =$ [ ]

$10 - 2 =$ [ ]

$8 + 8 =$ [ ]

$10 \div 2 =$ [ ]

$8 \times 2 =$ [ ]

$7 - 4 =$ [ ]

$12 \div 9 =$ [ ] REMAINDER [ ]

$0 \times 9 =$ [ ]

$14 - 8 =$ [ ]

$1 \times 1 =$ [ ]

$2 + 1 =$ [ ]

$8 + 1 =$ [ ]

$3 + 8 =$ [ ]

$15 \div 3 =$ [ ]

$3 \times 5 =$ [ ]

$9 + 4 =$ [ ]

$7 \times 6 =$ [ ]

$15 \div 4 =$ [ ] REMAINDER [ ]

$14 \div 2 =$ [ ]

$6 - 3 =$ [ ]

$6 \times 3 =$ [ ]

$4 \times 6 =$ [ ]

$14 - 5 =$ [ ]

$19 \div 9 =$ [ ] REMAINDER [ ]

$8 - 3 =$ [ ]

$3 + 3 =$ [ ]

End Time [ ] : [ ]

Start Time ☐ : ☐

$9 \times 7 =$ ☐

$4 + 3 =$ ☐

$2 + 6 =$ ☐

$8 - 6 =$ ☐

$9 - 5 =$ ☐

$6 \times 1 =$ ☐

$17 \div 8 =$ ☐ REMAINDER ☐

$4 + 2 =$ ☐

$12 - 6 =$ ☐

$2 \div 2 =$ ☐

$6 \times 7 =$ ☐

$19 \div 8 =$ ☐ REMAINDER ☐

$2 + 8 =$ ☐

$9 \times 4 =$ ☐

$7 - 4 =$ ☐

$7 \div 7 =$ ☐

$0 \times 7 =$ ☐

$16 \div 4 =$ ☐

$10 - 7 =$ ☐

$9 + 9 =$ ☐

$5 \times 5 =$ ☐

$19 \div 7 =$ ☐ REMAINDER ☐

$6 - 3 =$ ☐

$1 + 1 =$ ☐

$5 - 3 =$ ☐

$7 + 4 =$ ☐

$5 \times 8 =$ ☐

$1 \times 4 =$ ☐

$12 \div 3 =$ ☐

$3 + 1 =$ ☐

$15 - 9 =$ ☐

$14 \div 7 =$ ☐

$6 + 5 =$ ☐

$7 \div 6 =$ ☐ REMAINDER ☐

$2 \times 5 =$ ☐

$16 - 7 =$ ☐

$5 + 6 =$ ☐

$11 - 8 =$ ☐

$8 \times 2 =$ ☐

$14 \div 8 =$ ☐ REMAINDER ☐

$8 - 8 =$ ☐

$3 \times 7 =$ ☐

$9 + 7 =$ ☐

$18 \div 9 =$ ☐

$1 + 8 =$ ☐

$6 + 2 =$ ☐

$3 \times 5 =$ ☐

$5 - 1 =$ ☐

$10 - 5 =$ ☐

$11 \div 4 =$ ☐ REMAINDER ☐

$2 \div 1 =$ [ ]

$11 - 4 =$ [ ]

$9 + 5 =$ [ ]

$2 \times 8 =$ [ ]

$17 \div 4 =$ [ ] REMAINDER [ ]

$7 \times 1 =$ [ ]

$5 + 4 =$ [ ]

$12 - 8 =$ [ ]

$7 \times 3 =$ [ ]

$8 \div 1 =$ [ ]

$2 + 9 =$ [ ]

$8 \times 9 =$ [ ]

$18 \div 2 =$ [ ]

$3 - 3 =$ [ ]

$9 \times 5 =$ [ ]

$12 - 5 =$ [ ]

$1 + 7 =$ [ ]

$6 - 1 =$ [ ]

$4 + 1 =$ [ ]

$8 \div 5 =$ [ ] REMAINDER [ ]

$2 + 5 =$ [ ]

$9 + 2 =$ [ ]

$18 \div 4 =$ [ ] REMAINDER [ ]

$13 \div 8 =$ [ ] REMAINDER [ ]

$3 - 2 =$ [ ]

$9 - 7 =$ [ ]

$10 \div 5 =$ [ ]

$1 \times 8 =$ [ ]

$9 \times 3 =$ [ ]

$4 + 6 =$ [ ]

$6 \div 3 =$ [ ]

$2 \times 9 =$ [ ]

$3 \times 1 =$ [ ]

$14 - 8 =$ [ ]

$2 - 1 =$ [ ]

$10 \div 9 =$ [ ] REMAINDER [ ]

$4 \times 7 =$ [ ]

$10 - 6 =$ [ ]

$6 + 9 =$ [ ]

$8 + 5 =$ [ ]

$19 \div 2 =$ [ ] REMAINDER [ ]

$11 - 5 =$ [ ]

$5 \times 7 =$ [ ]

$6 + 1 =$ [ ]

$1 \times 3 =$ [ ]

$16 \div 8 =$ [ ]

$6 + 3 =$ [ ]

$6 - 2 =$ [ ]

$9 \times 8 =$ [ ]

$15 \div 5 =$ [ ]

End Time [ ] : [ ]

Start Time ☐ : ☐

$5 + 1 =$ ☐

$6 \times 8 =$ ☐

$7 - 5 =$ ☐

$5 \div 5 =$ ☐

$9 \times 2 =$ ☐

$14 - 9 =$ ☐

$9 \div 3 =$ ☐

$8 + 1 =$ ☐

$19 \div 6 =$ ☐ REMAINDER ☐

$13 - 4 =$ ☐

$4 \times 5 =$ ☐

$9 - 3 =$ ☐

$18 \div 6 =$ ☐

$3 + 3 =$ ☐

$9 \times 6 =$ ☐

$5 + 3 =$ ☐

$6 + 8 =$ ☐

$11 - 7 =$ ☐

$8 \times 5 =$ ☐

$7 + 8 =$ ☐

$11 - 9 =$ ☐

$14 \div 6 =$ ☐ REMAINDER ☐

$16 \div 2 =$ ☐

$2 \times 6 =$ ☐

$18 \div 3 =$ ☐

$18 - 9 =$ ☐

$5 \times 8 =$ ☐

$2 - 2 =$ ☐

$6 + 6 =$ ☐

$18 \div 8 =$ ☐ REMAINDER ☐

$13 - 7 =$ ☐

$3 + 7 =$ ☐

$5 \times 9 =$ ☐

$4 \times 2 =$ ☐

$7 + 7 =$ ☐

$19 \div 9 =$ ☐ REMAINDER ☐

$0 + 5 =$ ☐

$4 \times 4 =$ ☐

$5 - 2 =$ ☐

$4 + 8 =$ ☐

$13 \div 6 =$ ☐ REMAINDER ☐

$8 - 4 =$ ☐

$6 \times 5 =$ ☐

$2 + 7 =$ ☐

$7 - 3 =$ ☐

$13 \div 5 =$ ☐ REMAINDER ☐

$12 \div 4 =$ ☐

$8 - 2 =$ ☐

$3 \times 4 =$ ☐

$2 + 2 =$ ☐

$1 \times 6 =$ ☐

$13 - 6 =$ ☐

$3 \times 8 =$ ☐

$8 \div 8 =$ ☐

$5 + 7 =$ ☐

$6 \div 6 =$ ☐

$5 \times 2 =$ ☐

$12 - 4 =$ ☐

$5 + 8 =$ ☐

$9 \div 5 =$ ☐ REMAINDER ☐

$8 + 7 =$ ☐

$3 \times 2 =$ ☐

$9 - 6 =$ ☐

$5 + 5 =$ ☐

$8 \div 2 =$ ☐

$9 \times 0 =$ ☐

$15 - 6 =$ ☐

$7 \times 6 =$ ☐

$1 + 6 =$ ☐

$11 - 3 =$ ☐

$1 + 5 =$ ☐

$3 \times 6 =$ ☐

$11 - 2 =$ ☐

$13 \div 2 =$ ☐ REMAINDER ☐

$9 + 3 =$ ☐

$6 \times 4 =$ ☐

$15 - 7 =$ ☐

$12 \div 2 =$ ☐

$7 \times 9 =$ ☐

$3 + 2 =$ ☐

$8 - 1 =$ ☐

$5 \div 3 =$ ☐ REMAINDER ☐

$8 \div 4 =$ ☐

$4 \times 6 =$ ☐

$15 \div 8 =$ ☐ REMAINDER ☐

$4 - 2 =$ ☐

$9 \div 9 =$ ☐

$4 + 4 =$ ☐

$4 \times 8 =$ ☐

$13 \div 4 =$ ☐ REMAINDER ☐

$7 + 2 =$ ☐

$3 + 5 =$ ☐

$3 - 2 =$ ☐

$7 \times 3 =$ ☐

$15 \div 3 =$ ☐

$8 - 5 =$ ☐

$7 \times 8 =$ ☐

$6 \div 2 =$ ☐

$8 + 4 =$ ☐

$7 \div 5 =$ ☐ REMAINDER ☐

End Time ☐ : ☐

Start Time ☐ : ☐

$5 + 9 =$ ☐

$8 - 8 =$ ☐

$5 \times 8 =$ ☐

$5 \div 4 =$ ☐ REMAINDER ☐

$2 \times 4 =$ ☐

$7 + 9 =$ ☐

$10 - 7 =$ ☐

$1 \div 1 =$ ☐

$12 \div 6 =$ ☐

$7 \times 7 =$ ☐

$11 - 5 =$ ☐

$1 + 2 =$ ☐

$15 \div 8 =$ ☐ REMAINDER ☐

$9 \times 8 =$ ☐

$11 - 7 =$ ☐

$8 + 3 =$ ☐

$18 \div 2 =$ ☐

$9 - 2 =$ ☐

$8 \times 4 =$ ☐

$1 + 7 =$ ☐

$10 \div 2 =$ ☐

$4 + 2 =$ ☐

$15 \div 6 =$ ☐ REMAINDER ☐

$9 + 6 =$ ☐

$7 - 1 =$ ☐

$6 \times 5 =$ ☐

$9 - 3 =$ ☐

$1 + 3 =$ ☐

$3 + 2 =$ ☐

$5 \times 5 =$ ☐

$10 \div 6 =$ ☐ REMAINDER ☐

$8 \times 6 =$ ☐

$11 - 8 =$ ☐

$12 - 5 =$ ☐

$8 + 6 =$ ☐

$8 \times 5 =$ ☐

$6 \div 3 =$ ☐

$8 - 7 =$ ☐

$2 + 4 =$ ☐

$3 \times 8 =$ ☐

$14 - 7 =$ ☐

$6 - 3 =$ ☐

$8 + 2 =$ ☐

$13 \div 3 =$ ☐ REMAINDER ☐

$6 \div 6 =$ ☐

$3 \times 7 =$ ☐

$3 - 2 =$ ☐

$6 + 2 =$ ☐

$5 \times 3 =$ ☐

$18 \div 5 =$ ☐ REMAINDER ☐

$7 + 8 =$ ☐ $1 \times 1 =$ ☐ $18 \div 6 =$ ☐

$17 \div 8 =$ ☐ REMAINDER ☐ $9 + 5 =$ ☐ $16 - 8 =$ ☐

$3 + 3 =$ ☐ $15 - 8 =$ ☐ $4 \div 1 =$ ☐

$2 \times 2 =$ ☐ $18 \div 4 =$ ☐ REMAINDER ☐ $2 - 1 =$ ☐

$19 \div 6 =$ ☐ REMAINDER ☐ $7 \times 3 =$ ☐ $17 - 8 =$ ☐

$6 \times 6 =$ ☐ $15 - 6 =$ ☐ $3 + 6 =$ ☐

$18 - 9 =$ ☐ $9 - 0 =$ ☐ $3 \times 2 =$ ☐

$6 \times 7 =$ ☐ $1 \times 4 =$ ☐ $4 \times 9 =$ ☐

$4 + 7 =$ ☐ $18 \div 3 =$ ☐ $14 \div 2 =$ ☐

$5 - 4 =$ ☐ $5 + 4 =$ ☐ $5 - 5 =$ ☐

$2 \times 8 =$ ☐ $1 + 1 =$ ☐ $3 + 9 =$ ☐

$4 + 1 =$ ☐ $11 \div 5 =$ ☐ REMAINDER ☐ $14 - 6 =$ ☐

$13 \div 7 =$ ☐ REMAINDER ☐ $4 + 9 =$ ☐ $4 \div 2 =$ ☐

$8 - 4 =$ ☐ $8 \div 4 =$ ☐ $8 \times 3 =$ ☐

$7 \times 8 =$ ☐ $18 \div 9 =$ ☐ $6 + 7 =$ ☐

$9 \div 4 =$ ☐ REMAINDER ☐ $1 \times 5 =$ ☐ $8 \times 8 =$ ☐

$1 + 8 =$ ☐ $4 - 3 =$ ☐

End Time ☐ : ☐

Date ☐ M ☐ D

Start Time ☐ : ☐

$4 \times 4 =$ ☐

$14 \div 7 =$ ☐

$6 + 8 =$ ☐

$7 - 7 =$ ☐

$2 + 3 =$ ☐

$6 \times 2 =$ ☐

$14 - 5 =$ ☐

$9 \times 5 =$ ☐

$5 \div 5 =$ ☐

$9 - 5 =$ ☐

$5 + 1 =$ ☐

$11 \div 2 =$ ☐ REMAINDER ☐

$9 - 6 =$ ☐

$9 \times 3 =$ ☐

$3 + 1 =$ ☐

$12 \div 8 =$ ☐ REMAINDER ☐

$0 \times 5 =$ ☐

$4 - 2 =$ ☐

$8 \div 1 =$ ☐

$1 + 8 =$ ☐

$9 \div 2 =$ ☐ REMAINDER ☐

$8 + 5 =$ ☐

$14 - 8 =$ ☐

$15 \div 3 =$ ☐

$2 + 1 =$ ☐

$7 \times 4 =$ ☐

$12 - 7 =$ ☐

$3 \times 9 =$ ☐

$7 - 6 =$ ☐

$9 + 7 =$ ☐

$1 \times 2 =$ ☐

$4 + 5 =$ ☐

$16 \div 3 =$ ☐ REMAINDER ☐

$11 - 2 =$ ☐

$15 \div 5 =$ ☐

$7 - 5 =$ ☐

$2 \times 5 =$ ☐

$5 \times 7 =$ ☐

$7 + 7 =$ ☐

$14 \div 6 =$ ☐ REMAINDER ☐

$1 \times 9 =$ ☐

$7 + 4 =$ ☐

$12 \div 3 =$ ☐

$6 - 2 =$ ☐

$3 + 2 =$ ☐

$12 - 8 =$ ☐

$10 \div 7 =$ ☐ REMAINDER ☐

$9 + 1 =$ ☐

$12 - 4 =$ ☐

$3 \times 3 =$ ☐

$9 + 8 =$ ☐

$4 \times 5 =$ ☐

$15 - 7 =$ ☐

$3 \div 3 =$ ☐

$2 + 8 =$ ☐

$9 \div 9 =$ ☐

$3 + 5 =$ ☐

$17 - 9 =$ ☐

$1 \times 3 =$ ☐

$5 - 2 =$ ☐

$9 - 0 =$ ☐

$5 \div 2 =$ ☐ REMAINDER ☐

$6 \times 8 =$ ☐

$5 \div 1 =$ ☐

$9 \times 4 =$ ☐

$2 + 7 =$ ☐

$6 \times 9 =$ ☐

$3 \times 4 =$ ☐

$19 \div 7 =$ ☐ REMAINDER ☐

$13 - 5 =$ ☐

$5 + 3 =$ ☐

$4 - 1 =$ ☐

$6 \times 4 =$ ☐

$10 \div 3 =$ ☐ REMAINDER ☐

$5 + 6 =$ ☐

$8 \times 9 =$ ☐

$14 \div 5 =$ ☐ REMAINDER ☐

$5 + 7 =$ ☐

$13 - 7 =$ ☐

$13 \div 9 =$ ☐ REMAINDER ☐

$9 - 9 =$ ☐

$16 \div 2 =$ ☐

$7 + 2 =$ ☐

$3 \times 6 =$ ☐

$11 - 3 =$ ☐

$1 + 9 =$ ☐

$2 \times 6 =$ ☐

$10 \div 5 =$ ☐

$12 \div 7 =$ ☐ REMAINDER ☐

$7 + 1 =$ ☐

$3 \times 5 =$ ☐

$7 - 2 =$ ☐

$12 \div 2 =$ ☐

$2 - 2 =$ ☐

$4 \times 8 =$ ☐

$3 + 4 =$ ☐

$9 \div 3 =$ ☐

$2 \times 1 =$ ☐

$5 + 5 =$ ☐

$14 - 9 =$ ☐

End Time ☐ : ☐

I. Counting Test

Measure the time required for you to count from 1 to 120 aloud as fast as you can.

☐ sec.

II. Word Memorization Test

Memorize as many words as you can **within two minutes.**

hoof	oil	glue	peanut	trial	fence
spring	credit	mystery	desk	fish	ewer
camel	adult	embark	sensation	Tuna	cloud
slope	overcoat	kelp	ruler	tuna	carousel
automatic	chemistry	order	field	symbol	violet

Write out as many words as you can remember **in two minutes** on the back of this page. How many words can you remember?

Number of words memorized ☐ words

Word Memorization Test Answers

III. Stroop Test

Please take the **Stroop Test** for Week 8, located on page **xi** of the Appendix.

Start Time ☐ : ☐

$8 \div 5 =$ ☐ REMAINDER ☐

$4 + 2 =$ ☐

$14 \div 2 =$ ☐

$15 - 9 =$ ☐

$9 \times 4 =$ ☐

$4 + 5 =$ ☐

$8 - 2 =$ ☐

$6 + 6 =$ ☐

$17 \div 9 =$ ☐ REMAINDER ☐

$5 \times 9 =$ ☐

$6 - 3 =$ ☐

$4 \times 5 =$ ☐

$6 \div 4 =$ ☐ REMAINDER ☐

$9 - 9 =$ ☐

$8 \times 3 =$ ☐

$8 + 9 =$ ☐

$12 \div 2 =$ ☐

$7 - 5 =$ ☐

$8 \times 9 =$ ☐

$7 + 6 =$ ☐

$16 - 9 =$ ☐

$9 + 9 =$ ☐

$4 \times 9 =$ ☐

$2 + 4 =$ ☐

$8 \div 6 =$ ☐ REMAINDER ☐

$5 \times 6 =$ ☐

$14 \div 7 =$ ☐

$8 - 5 =$ ☐

$1 + 7 =$ ☐

$4 + 6 =$ ☐

$8 - 4 =$ ☐

$7 \div 5 =$ ☐ REMAINDER ☐

$9 \times 3 =$ ☐

$14 - 8 =$ ☐

$8 + 8 =$ ☐

$6 - 4 =$ ☐

$3 \times 8 =$ ☐

$13 \div 2 =$ ☐ REMAINDER ☐

$3 + 5 =$ ☐

$2 + 2 =$ ☐

$10 \div 2 =$ ☐

$8 \times 1 =$ ☐

$15 \div 5 =$ ☐

$11 - 7 =$ ☐

$6 \times 6 =$ ☐

$12 - 9 =$ ☐

$1 \times 0 =$ ☐

$9 \div 3 =$ ☐

$13 - 8 =$ ☐

$3 + 3 =$ ☐

$7 \div 7 =$ [ ]

$0 \times 3 =$ [ ]

$9 + 8 =$ [ ]

$14 - 7 =$ [ ]

$7 \times 9 =$ [ ]

$2 + 6 =$ [ ]

$9 \times 9 =$ [ ]

$9 \div 1 =$ [ ]

$10 \div 5 =$ [ ]

$8 - 7 =$ [ ]

$5 \times 1 =$ [ ]

$7 + 1 =$ [ ]

$3 \div 2 =$ [ ] REMAINDER [ ]

$9 - 2 =$ [ ]

$2 \times 5 =$ [ ]

$12 - 7 =$ [ ]

$1 + 8 =$ [ ]

$9 + 4 =$ [ ]

$4 \div 2 =$ [ ]

$10 - 7 =$ [ ]

$1 + 9 =$ [ ]

$11 - 3 =$ [ ]

$2 \times 4 =$ [ ]

$15 \div 9 =$ [ ] REMAINDER [ ]

$8 - 1 =$ [ ]

$9 \times 7 =$ [ ]

$8 \div 4 =$ [ ]

$4 \times 3 =$ [ ]

$5 \div 2 =$ [ ] REMAINDER [ ]

$2 + 1 =$ [ ]

$10 \div 3 =$ [ ] REMAINDER [ ]

$9 \times 2 =$ [ ]

$2 - 1 =$ [ ]

$2 + 7 =$ [ ]

$3 \times 3 =$ [ ]

$9 + 7 =$ [ ]

$8 \times 5 =$ [ ]

$11 - 4 =$ [ ]

$11 \div 6 =$ [ ] REMAINDER [ ]

$1 - 1 =$ [ ]

$12 - 8 =$ [ ]

$16 \div 8 =$ [ ]

$2 + 5 =$ [ ]

$19 \div 2 =$ [ ] REMAINDER [ ]

$6 \times 8 =$ [ ]

$8 + 5 =$ [ ]

$16 \div 4 =$ [ ]

$9 + 1 =$ [ ]

$9 \times 6 =$ [ ]

$5 - 4 =$ [ ]

End Time [ ] : [ ]

Date ☐ M ☐ D

Start Time ☐ : ☐

$8 \times 4 =$ ☐

$16 \div 2 =$ ☐

$5 + 9 =$ ☐

$3 - 3 =$ ☐

$1 \times 8 =$ ☐

$18 \div 9 =$ ☐

$8 \times 2 =$ ☐

$10 - 9 =$ ☐

$14 \div 8 =$ ☐ REMAINDER ☐

$2 \times 9 =$ ☐

$5 + 8 =$ ☐

$3 + 2 =$ ☐

$9 - 7 =$ ☐

$3 - 1 =$ ☐

$1 + 5 =$ ☐

$16 \div 9 =$ ☐ REMAINDER ☐

$5 + 2 =$ ☐

$8 \times 1 =$ ☐

$15 - 6 =$ ☐

$0 + 8 =$ ☐

$12 \div 4 =$ ☐

$10 - 8 =$ ☐

$7 + 4 =$ ☐

$2 \times 6 =$ ☐

$5 \times 7 =$ ☐

$19 \div 3 =$ ☐ REMAINDER ☐

$9 - 3 =$ ☐

$15 \div 8 =$ ☐ REMAINDER ☐

$7 + 5 =$ ☐

$19 \div 8 =$ ☐ REMAINDER ☐

$10 - 5 =$ ☐

$3 + 1 =$ ☐

$7 \times 5 =$ ☐

$8 - 8 =$ ☐

$5 + 3 =$ ☐

$8 \div 8 =$ ☐

$6 \times 7 =$ ☐

$7 - 4 =$ ☐

$14 \div 4 =$ ☐ REMAINDER ☐

$7 \times 7 =$ ☐

$9 + 6 =$ ☐

$4 \times 7 =$ ☐

$6 + 7 =$ ☐

$11 - 5 =$ ☐

$6 \div 3 =$ ☐

$12 - 5 =$ ☐

$6 + 2 =$ ☐

$7 \times 1 =$ ☐

$6 - 5 =$ ☐

$12 \div 2 =$ ☐

$4 + 1 =$ [ ]

$14 - 9 =$ [ ]

$2 \times 0 =$ [ ]

$18 \div 3 =$ [ ]

$1 + 4 =$ [ ]

$3 \times 2 =$ [ ]

$6 - 2 =$ [ ]

$4 + 3 =$ [ ]

$6 \times 4 =$ [ ]

$17 \div 3 =$ [ ] REMAINDER [ ]

$4 - 3 =$ [ ]

$5 - 2 =$ [ ]

$5 \times 3 =$ [ ]

$18 \div 2 =$ [ ]

$8 + 3 =$ [ ]

$15 \div 7 =$ [ ] REMAINDER [ ]

$4 \times 6 =$ [ ]

$13 - 4 =$ [ ]

$2 \times 8 =$ [ ]

$6 + 3 =$ [ ]

$16 \div 5 =$ [ ] REMAINDER [ ]

$3 \div 1 =$ [ ]

$10 - 6 =$ [ ]

$2 + 9 =$ [ ]

$15 \div 6 =$ [ ] REMAINDER [ ]

$9 \times 1 =$ [ ]

$9 - 2 =$ [ ]

$6 \times 3 =$ [ ]

$8 \div 2 =$ [ ]

$15 \div 3 =$ [ ]

$1 + 1 =$ [ ]

$9 + 3 =$ [ ]

$1 \times 9 =$ [ ]

$12 - 3 =$ [ ]

$19 \div 7 =$ [ ] REMAINDER [ ]

$6 + 9 =$ [ ]

$4 \times 8 =$ [ ]

$12 - 6 =$ [ ]

$5 + 6 =$ [ ]

$6 \times 9 =$ [ ]

$12 \div 3 =$ [ ]

$13 - 9 =$ [ ]

$9 + 0 =$ [ ]

$13 \div 3 =$ [ ] REMAINDER [ ]

$3 - 2 =$ [ ]

$8 + 2 =$ [ ]

$9 - 6 =$ [ ]

$1 \times 6 =$ [ ]

$9 \times 5 =$ [ ]

$18 \div 6 =$ [ ]

End Time [ ] : [ ]

Start Time ☐ : ☐

$7 + 4 =$ ☐

$5 \times 2 =$ ☐

$9 \div 5 =$ ☐ REMAINDER ☐

$14 - 9 =$ ☐

$5 + 3 =$ ☐

$2 \times 3 =$ ☐

$4 \div 2 =$ ☐

$9 - 7 =$ ☐

$8 \div 5 =$ ☐ REMAINDER ☐

$9 - 1 =$ ☐

$2 + 6 =$ ☐

$9 \times 1 =$ ☐

$9 + 4 =$ ☐

$4 - 1 =$ ☐

$7 \times 2 =$ ☐

$12 \div 2 =$ ☐

$9 + 5 =$ ☐

$5 \times 3 =$ ☐

$16 \div 5 =$ ☐ REMAINDER ☐

$5 - 2 =$ ☐

$6 + 3 =$ ☐

$3 \times 1 =$ ☐

$6 - 5 =$ ☐

$13 - 4 =$ ☐

$6 \div 6 =$ ☐

$7 \times 4 =$ ☐

$16 \div 2 =$ ☐

$5 + 7 =$ ☐

$11 - 8 =$ ☐

$0 + 5 =$ ☐

$3 + 3 =$ ☐

$15 \div 5 =$ ☐

$9 \times 2 =$ ☐

$7 + 1 =$ ☐

$12 \div 8 =$ ☐ REMAINDER ☐

$12 - 4 =$ ☐

$6 \times 9 =$ ☐

$2 - 0 =$ ☐

$5 + 8 =$ ☐

$15 \div 2 =$ ☐ REMAINDER ☐

$4 + 1 =$ ☐

$11 - 6 =$ ☐

$14 \div 2 =$ ☐

$3 + 8 =$ ☐

$7 \times 1 =$ ☐

$5 \times 8 =$ ☐

$16 \div 7 =$ ☐ REMAINDER ☐

$9 \times 9 =$ ☐

$14 - 5 =$ ☐

$9 - 9 =$ ☐

$19 \div 5 =$ ☐ REMAINDER ☐　　$5 + 9 =$ ☐　　　　$8 \div 8 =$ ☐

$4 + 2 =$ ☐　　　　　　　　$6 \times 7 =$ ☐　　　　$6 \times 4 =$ ☐

$1 + 6 =$ ☐　　　　　　　　$2 + 5 =$ ☐　　　　$9 + 1 =$ ☐

$13 - 6 =$ ☐　　　　　　　$19 \div 8 =$ ☐ REMAINDER ☐　　$5 \times 7 =$ ☐

$12 \div 3 =$ ☐　　　　　　$14 - 6 =$ ☐　　　　$7 - 1 =$ ☐

$3 \times 2 =$ ☐　　　　　　　$1 \times 2 =$ ☐　　　　$8 \div 1 =$ ☐

$10 - 9 =$ ☐　　　　　　　$11 \div 7 =$ ☐ REMAINDER ☐　　$7 + 9 =$ ☐

$3 + 5 =$ ☐　　　　　　　　$4 - 4 =$ ☐　　　　$2 - 1 =$ ☐

$5 \times 9 =$ ☐　　　　　　　$11 - 7 =$ ☐　　　　$4 \times 3 =$ ☐

$3 \times 3 =$ ☐　　　　　　　$18 \div 3 =$ ☐　　　　$4 + 9 =$ ☐

$11 \div 8 =$ ☐ REMAINDER ☐　　$5 \times 1 =$ ☐　　　　$18 \div 4 =$ ☐ REMAINDER ☐

$8 \times 7 =$ ☐　　　　　　　$9 + 9 =$ ☐　　　　$12 - 9 =$ ☐

$9 - 4 =$ ☐　　　　　　　　$6 \div 2 =$ ☐　　　　$0 \times 7 =$ ☐

$6 + 9 =$ ☐　　　　　　　　$5 - 4 =$ ☐　　　　$6 - 3 =$ ☐

$12 - 8 =$ ☐　　　　　　　$4 \times 2 =$ ☐　　　　$3 + 4 =$ ☐

$10 \div 3 =$ ☐ REMAINDER ☐　　$1 + 7 =$ ☐　　　　$15 \div 3 =$ ☐

$3 \times 4 =$ ☐　　　　　　　$12 \div 6 =$ ☐　　　　**End Time** ☐ : ☐

Date ☐ M ☐ D

Start Time ☐ : ☐

$8 - 2 =$ ☐ $2 + 4 =$ ☐

$8 \times 3 =$ ☐ $12 - 5 =$ ☐ $3 \times 5 =$ ☐

$12 \div 7 =$ ☐ REMAINDER ☐ $6 \times 2 =$ ☐ $4 - 2 =$ ☐

$10 - 3 =$ ☐ $9 \div 6 =$ ☐ REMAINDER ☐ $9 \div 3 =$ ☐

$6 \div 5 =$ ☐ REMAINDER ☐ $9 \times 7 =$ ☐ $1 \times 7 =$ ☐

$7 - 3 =$ ☐ $7 - 4 =$ ☐ $4 + 8 =$ ☐

$9 + 3 =$ ☐ $1 + 8 =$ ☐ $6 - 2 =$ ☐

$13 - 5 =$ ☐ $9 \times 3 =$ ☐ $2 \div 2 =$ ☐

$6 + 2 =$ ☐ $8 + 8 =$ ☐ $13 - 7 =$ ☐

$14 \div 8 =$ ☐ REMAINDER ☐ $16 \div 4 =$ ☐ $1 + 1 =$ ☐

$2 \times 4 =$ ☐ $11 - 3 =$ ☐ $4 + 4 =$ ☐

$3 - 2 =$ ☐ $3 + 5 =$ ☐ $5 \div 5 =$ ☐

$9 + 7 =$ ☐ $6 - 4 =$ ☐ $1 \times 1 =$ ☐

$1 \times 6 =$ ☐ $12 \div 4 =$ ☐ $4 + 6 =$ ☐

$5 + 1 =$ ☐ $8 + 4 =$ ☐ $8 \div 7 =$ ☐ REMAINDER ☐

$13 \div 9 =$ ☐ REMAINDER ☐ $8 \times 9 =$ ☐ $5 \times 5 =$ ☐

$7 \times 3 =$ ☐ $18 \div 6 =$ ☐ $13 - 8 =$ ☐

$6 + 1 =$ ☐ $5 + 6 =$ ☐ $9 \times 6 =$ ☐

$12 - 3 =$ ☐ $2 \times 7 =$ ☐ $13 - 9 =$ ☐

$18 \div 7 =$ ☐ REMAINDER ☐ $8 + 1 =$ ☐ $6 + 5 =$ ☐

$3 + 2 =$ ☐ $6 \div 3 =$ ☐ $18 \div 9 =$ ☐

$1 \times 3 =$ ☐ $10 - 7 =$ ☐ $2 \times 8 =$ ☐

$14 \div 7 =$ ☐ $3 \times 9 =$ ☐ $4 - 4 =$ ☐

$0 \times 2 =$ ☐ $16 \div 8 =$ ☐ $8 + 9 =$ ☐

$2 + 1 =$ ☐ $3 + 9 =$ ☐ $13 \div 4 =$ ☐ REMAINDER ☐

$9 - 8 =$ ☐ $4 \times 6 =$ ☐ $7 + 5 =$ ☐

$3 \div 3 =$ ☐ $7 \div 1 =$ ☐ $5 \times 4 =$ ☐

$8 + 2 =$ ☐ $5 - 5 =$ ☐ $15 \div 9 =$ ☐ REMAINDER ☐

$14 \div 5 =$ ☐ REMAINDER ☐ $7 \times 7 =$ ☐ $8 - 6 =$ ☐

$6 \times 5 =$ ☐ $15 \div 8 =$ ☐ REMAINDER ☐ $2 \times 6 =$ ☐

$9 - 3 =$ ☐ $15 - 6 =$ ☐ $7 + 2 =$ ☐

$5 \div 4 =$ ☐ REMAINDER ☐ $2 + 7 =$ ☐ $17 - 8 =$ ☐

$3 - 0 =$ ☐ $16 - 9 =$ ☐ $8 \div 4 =$ ☐

$8 \times 5 =$ ☐ $7 \times 6 =$ ☐   End Time ☐ : ☐

Start Time ☐ : ☐

$2 + 4 =$ ☐

$17 \div 5 =$ ☐ REMAINDER ☐

$9 - 3 =$ ☐

$5 \times 4 =$ ☐

$9 + 3 =$ ☐

$17 - 9 =$ ☐

$4 \times 0 =$ ☐

$18 \div 8 =$ ☐ REMAINDER ☐

$5 \times 2 =$ ☐

$4 - 0 =$ ☐

$4 + 2 =$ ☐

$1 + 3 =$ ☐

$15 \div 3 =$ ☐

$8 - 4 =$ ☐

$8 \times 1 =$ ☐

$10 \div 7 =$ ☐ REMAINDER ☐

$12 \div 4 =$ ☐

$5 \times 9 =$ ☐

$12 - 4 =$ ☐

$2 + 7 =$ ☐

$1 \times 9 =$ ☐

$10 - 8 =$ ☐

$9 - 5 =$ ☐

$5 \times 1 =$ ☐

$13 \div 2 =$ ☐ REMAINDER ☐

$10 - 2 =$ ☐

$1 + 8 =$ ☐

$6 \times 4 =$ ☐

$18 \div 5 =$ ☐ REMAINDER ☐

$10 - 1 =$ ☐

$7 + 6 =$ ☐

$2 \div 2 =$ ☐

$3 + 9 =$ ☐

$6 \times 3 =$ ☐

$9 - 6 =$ ☐

$9 + 1 =$ ☐

$16 \div 4 =$ ☐

$4 \times 8 =$ ☐

$6 - 1 =$ ☐

$5 + 9 =$ ☐

$10 \div 2 =$ ☐

$2 + 1 =$ ☐

$1 \times 0 =$ ☐

$7 \div 1 =$ ☐

$16 - 9 =$ ☐

$3 \div 2 =$ ☐ REMAINDER ☐

$4 \times 6 =$ ☐

$8 + 8 =$ ☐

$6 - 5 =$ ☐

$2 + 5 =$ ☐

Time Required [ ] : [ ]

$5 \times 3 =$ [ ] $4 \times 2 =$ [ ] $17 - 8 =$ [ ]

$12 \div 6 =$ [ ] $11 - 6 =$ [ ] $6 \div 6 =$ [ ]

$2 - 1 =$ [ ] $8 \times 8 =$ [ ] $4 \times 7 =$ [ ]

$17 \div 2 =$ [ ] REMAINDER [ ] $1 \div 1 =$ [ ] $8 + 5 =$ [ ]

$3 \times 8 =$ [ ] $6 + 2 =$ [ ] $0 + 6 =$ [ ]

$5 + 5 =$ [ ] $7 - 6 =$ [ ] $19 \div 8 =$ [ ] REMAINDER [ ]

$6 - 6 =$ [ ] $11 \div 5 =$ [ ] REMAINDER [ ] $9 + 7 =$ [ ]

$9 \times 2 =$ [ ] $18 \div 3 =$ [ ] $8 \times 5 =$ [ ]

$8 + 6 =$ [ ] $0 + 4 =$ [ ] $5 - 2 =$ [ ]

$17 \div 7 =$ [ ] REMAINDER [ ] $6 + 1 =$ [ ] $4 \div 2 =$ [ ]

$1 + 6 =$ [ ] $2 \times 9 =$ [ ] $4 \times 4 =$ [ ]

$2 \times 3 =$ [ ] $11 \div 4 =$ [ ] REMAINDER [ ] $7 - 2 =$ [ ]

$12 - 9 =$ [ ] $4 + 9 =$ [ ] $11 - 5 =$ [ ]

$7 \div 2 =$ [ ] REMAINDER [ ] $15 \div 5 =$ [ ] $10 \div 5 =$ [ ]

$15 - 6 =$ [ ] $8 - 1 =$ [ ] $8 + 1 =$ [ ]

$5 + 6 =$ [ ] $4 \times 5 =$ [ ] $9 \times 8 =$ [ ]

$7 \times 4 =$ [ ] $11 - 3 =$ [ ]   End Time [ ] : [ ]

I. Counting Test

Measure the time required for you to count from 1 to 120 aloud as fast as you can.

☐ sec.

II. Word Memorization Test

Memorize as many words as you can **within two minutes.**

doll	clay	gadget	politician	gourd	air
skylark	cemetery	pancake	yellow	princess	burn
mackerel	speed	naked	employ	pattern	factory
pride	parrot	jellyfish	illustration	electric	grape
deck	sparrow	debts	square	symbol	cartoon

Write out as many words as you can remember **in two minutes** on the back of this page. How many words can you remember?

Number of words memorized ☐ words

Word Memorization Test Answers

III. Stroop Test

Please take the **Stroop Test** for Week 9, located on page **xii** of the Appendix.

Date ☐ M ☐ D

Start Time ☐ : ☐

$8 + 3 =$ ☐

$16 \div 2 =$ ☐

$2 \times 7 =$ ☐

$13 - 8 =$ ☐

$3 + 3 =$ ☐

$10 \div 8 =$ ☐ REMAINDER ☐

$4 \div 1 =$ ☐

$18 - 9 =$ ☐

$12 - 7 =$ ☐

$5 + 0 =$ ☐

$6 \times 8 =$ ☐

$14 \div 8 =$ ☐ REMAINDER ☐

$6 \times 5 =$ ☐

$4 + 8 =$ ☐

$9 - 2 =$ ☐

$3 \times 5 =$ ☐

$7 + 5 =$ ☐

$14 - 6 =$ ☐

$18 \div 7 =$ ☐ REMAINDER ☐

$9 - 4 =$ ☐

$1 \times 5 =$ ☐

$5 + 3 =$ ☐

$2 - 2 =$ ☐

$17 \div 8 =$ ☐ REMAINDER ☐

$8 \times 3 =$ ☐

$6 + 3 =$ ☐

$5 \div 1 =$ ☐

$6 + 7 =$ ☐

$4 - 3 =$ ☐

$18 \div 9 =$ ☐

$7 \times 1 =$ ☐

$14 - 8 =$ ☐

$9 \times 0 =$ ☐

$2 \div 1 =$ ☐

$8 - 3 =$ ☐

$9 + 9 =$ ☐

$2 + 6 =$ ☐

$1 \times 3 =$ ☐

$5 - 4 =$ ☐

$16 \div 9 =$ ☐ REMAINDER ☐

$2 \times 2 =$ ☐

$13 - 9 =$ ☐

$8 + 9 =$ ☐

$6 \times 7 =$ ☐

$1 + 4 =$ ☐

$8 \div 2 =$ ☐

$4 - 1 =$ ☐

$3 \times 7 =$ ☐

$1 + 5 =$ ☐

$10 \div 3 =$ ☐ REMAINDER ☐

$11 - 7 =$ ☐

$3 + 2 =$ ☐

$14 \div 7 =$ ☐

$8 \times 9 =$ ☐

$4 + 3 =$ ☐

$8 + 7 =$ ☐

$17 \div 6 =$ ☐ REMAINDER ☐

$7 \times 8 =$ ☐

$5 - 3 =$ ☐

$18 \div 6 =$ ☐

$3 \times 2 =$ ☐

$14 - 5 =$ ☐

$7 \div 7 =$ ☐

$9 \times 4 =$ ☐

$5 + 7 =$ ☐

$7 - 6 =$ ☐

$10 \div 6 =$ ☐ REMAINDER ☐

$13 \div 8 =$ ☐ REMAINDER ☐

$7 \times 3 =$ ☐

$3 - 1 =$ ☐

$5 + 2 =$ ☐

$0 \times 5 =$ ☐

$3 + 4 =$ ☐

$4 + 7 =$ ☐

$8 \div 4 =$ ☐

$7 \times 7 =$ ☐

$10 - 4 =$ ☐

$16 \div 8 =$ ☐

$4 \times 1 =$ ☐

$13 - 7 =$ ☐

$9 + 6 =$ ☐

$9 \times 7 =$ ☐

$7 \div 3 =$ ☐ REMAINDER ☐

$13 - 6 =$ ☐

$4 + 5 =$ ☐

$12 \div 2 =$ ☐

$7 - 4 =$ ☐

$3 \times 9 =$ ☐

$14 \div 2 =$ ☐

$9 - 1 =$ ☐

$7 + 4 =$ ☐

$5 \times 7 =$ ☐

$14 \div 5 =$ ☐ REMAINDER ☐

$3 \times 3 =$ ☐

$6 - 6 =$ ☐

$3 + 6 =$ ☐

$10 - 7 =$ ☐

$7 \div 6 =$ ☐ REMAINDER ☐

$6 + 8 =$ ☐

$7 \times 5 =$ ☐

End Time ☐ : ☐

Date ☐ M ☐ D

Start Time ☐ : ☐

$4 + 8 =$ ☐

$12 - 3 =$ ☐

$13 \div 3 =$ ☐ REMAINDER ☐

$7 \times 7 =$ ☐

$3 + 6 =$ ☐

$12 \div 3 =$ ☐

$8 \times 1 =$ ☐

$15 - 9 =$ ☐

$10 \div 5 =$ ☐

$7 + 3 =$ ☐

$2 \times 7 =$ ☐

$11 - 8 =$ ☐

$5 + 4 =$ ☐

$8 \times 6 =$ ☐

$6 \div 2 =$ ☐

$12 - 9 =$ ☐

$4 + 4 =$ ☐

$4 - 3 =$ ☐

$2 \times 5 =$ ☐

$16 \div 2 =$ ☐

$5 - 3 =$ ☐

$3 + 4 =$ ☐

$7 - 6 =$ ☐

$6 + 3 =$ ☐

$7 \div 4 =$ ☐ REMAINDER ☐

$12 - 5 =$ ☐

$7 + 7 =$ ☐

$8 \times 4 =$ ☐

$11 \div 8 =$ ☐ REMAINDER ☐

$7 \times 4 =$ ☐

$9 - 1 =$ ☐

$16 \div 8 =$ ☐

$5 + 7 =$ ☐

$0 \times 5 =$ ☐

$10 - 4 =$ ☐

$9 \div 8 =$ ☐ REMAINDER ☐

$4 + 1 =$ ☐

$6 - 2 =$ ☐

$3 \times 5 =$ ☐

$9 \div 3 =$ ☐

$7 \times 9 =$ ☐

$4 \times 5 =$ ☐

$3 - 0 =$ ☐

$8 \div 5 =$ ☐ REMAINDER ☐

$5 + 1 =$ ☐

$6 + 5 =$ ☐

$7 \div 6 =$ ☐ REMAINDER ☐

$5 - 2 =$ ☐

$9 \times 9 =$ ☐

$7 + 5 =$ ☐

$19 \div 5 =$ ☐ REMAINDER ☐

$1 + 3 =$ ☐

$14 - 5 =$ ☐

$4 \times 3 =$ ☐

$9 + 5 =$ ☐

$2 - 1 =$ ☐

$12 \div 5 =$ ☐ REMAINDER ☐

$2 \times 8 =$ ☐

$7 + 2 =$ ☐

$15 \div 5 =$ ☐

$18 \div 4 =$ ☐ REMAINDER ☐

$11 - 5 =$ ☐

$2 \times 2 =$ ☐

$12 \div 6 =$ ☐

$9 \times 5 =$ ☐

$7 + 0 =$ ☐

$10 - 8 =$ ☐

$8 \times 8 =$ ☐

$16 \div 4 =$ ☐

$14 - 7 =$ ☐

$8 + 6 =$ ☐

$10 \div 4 =$ ☐ REMAINDER ☐

$4 + 5 =$ ☐

$18 \div 5 =$ ☐ REMAINDER ☐

$8 \times 5 =$ ☐

$9 \times 8 =$ ☐

$12 - 8 =$ ☐

$9 - 5 =$ ☐

$12 \div 4 =$ ☐

$4 - 0 =$ ☐

$3 + 9 =$ ☐

$6 \times 2 =$ ☐

$7 \times 6 =$ ☐

$5 + 8 =$ ☐

$18 - 9 =$ ☐

$13 \div 7 =$ ☐ REMAINDER ☐

$2 + 0 =$ ☐

$7 \times 2 =$ ☐

$5 + 9 =$ ☐

$15 \div 3 =$ ☐

$9 \times 1 =$ ☐

$6 - 4 =$ ☐

$4 \div 2 =$ ☐

$6 + 1 =$ ☐

$1 \times 9 =$ ☐

$7 + 9 =$ ☐

$3 - 3 =$ ☐

$6 \times 5 =$ ☐

$3 - 2 =$ ☐

$18 \div 6 =$ ☐

End Time ☐ : ☐

Day 48

Start Time ☐ : ☐

6 + 7 = ☐

3 + 1 = ☐

14 ÷ 6 = ☐ REMAINDER ☐

9 − 4 = ☐

4 + 3 = ☐

8 ÷ 2 = ☐

10 − 6 = ☐

19 ÷ 6 = ☐ REMAINDER ☐

9 × 4 = ☐

9 − 7 = ☐

11 − 3 = ☐

8 × 3 = ☐

4 + 2 = ☐

6 ÷ 6 = ☐

4 × 4 = ☐

6 × 7 = ☐

8 − 3 = ☐

6 + 2 = ☐

18 ÷ 9 = ☐

9 × 7 = ☐

4 − 2 = ☐

5 + 4 = ☐

3 − 1 = ☐

18 ÷ 3 = ☐

8 + 4 = ☐

5 × 9 = ☐

19 ÷ 3 = ☐ REMAINDER ☐

4 + 9 = ☐

13 ÷ 5 = ☐ REMAINDER ☐

2 × 0 = ☐

5 × 6 = ☐

6 + 9 = ☐

12 − 6 = ☐

17 − 9 = ☐

7 ÷ 1 = ☐

2 × 9 = ☐

14 ÷ 7 = ☐

2 + 6 = ☐

7 × 5 = ☐

1 + 2 = ☐

10 − 9 = ☐

4 × 1 = ☐

9 + 6 = ☐

7 − 7 = ☐

17 ÷ 7 = ☐ REMAINDER ☐

4 + 6 = ☐

14 ÷ 4 = ☐ REMAINDER ☐

4 × 9 = ☐

8 − 4 = ☐

11 − 7 = ☐

$2 + 8 =$ ⬚

$6 \times 8 =$ ⬚

$13 - 5 =$ ⬚

$7 \div 7 =$ ⬚

$1 + 8 =$ ⬚

$4 \times 7 =$ ⬚

$7 - 5 =$ ⬚

$1 \times 2 =$ ⬚

$7 + 8 =$ ⬚

$9 - 6 =$ ⬚

$4 \div 1 =$ ⬚

$5 \times 7 =$ ⬚

$9 \div 6 =$ ⬚ REMAINDER ⬚

$8 + 7 =$ ⬚

$2 \times 3 =$ ⬚

$12 \div 2 =$ ⬚

$8 - 6 =$ ⬚

$16 \div 7 =$ ⬚ REMAINDER ⬚

$4 - 0 =$ ⬚

$1 \times 6 =$ ⬚

$12 - 7 =$ ⬚

$6 \div 3 =$ ⬚

$3 + 5 =$ ⬚

$2 + 9 =$ ⬚

$4 \times 2 =$ ⬚

$13 - 8 =$ ⬚

$2 + 3 =$ ⬚

$6 \times 6 =$ ⬚

$16 \div 6 =$ ⬚ REMAINDER ⬚

$10 \div 2 =$ ⬚

$4 + 7 =$ ⬚

$6 \times 1 =$ ⬚

$14 - 6 =$ ⬚

$18 \div 8 =$ ⬚ REMAINDER ⬚

$8 + 1 =$ ⬚

$16 - 7 =$ ⬚

$8 \div 4 =$ ⬚

$19 \div 2 =$ ⬚ REMAINDER ⬚

$5 \times 4 =$ ⬚

$1 + 6 =$ ⬚

$14 \div 2 =$ ⬚

$15 - 8 =$ ⬚

$3 \times 8 =$ ⬚

$2 + 1 =$ ⬚

$6 - 4 =$ ⬚

$12 \div 9 =$ ⬚ REMAINDER ⬚

$6 \times 9 =$ ⬚

$1 + 9 =$ ⬚

$4 - 1 =$ ⬚

$8 \times 0 =$ ⬚

End Time ⬚ : ⬚

Date ☐ M ☐ D

Start Time ☐ : ☐

8 − 4 = ☐

2 × 6 = ☐

17 ÷ 2 = ☐ REMAINDER ☐

3 + 8 = ☐

17 − 9 = ☐

4 × 8 = ☐

9 + 8 = ☐

3 + 4 = ☐

6 − 4 = ☐

10 ÷ 5 = ☐

6 × 1 = ☐

9 + 2 = ☐

5 ÷ 2 = ☐ REMAINDER ☐

12 − 3 = ☐

15 ÷ 5 = ☐

1 × 8 = ☐

1 + 7 = ☐

5 × 4 = ☐

3 ÷ 2 = ☐ REMAINDER ☐

4 − 1 = ☐

9 × 7 = ☐

8 + 1 = ☐

6 ÷ 3 = ☐

12 − 9 = ☐

5 × 9 = ☐

5 − 1 = ☐

4 + 7 = ☐

17 ÷ 4 = ☐ REMAINDER ☐

4 × 6 = ☐

6 − 3 = ☐

11 − 9 = ☐

18 ÷ 5 = ☐ REMAINDER ☐

5 + 4 = ☐

8 × 9 = ☐

4 ÷ 2 = ☐

7 + 6 = ☐

10 − 2 = ☐

9 × 9 = ☐

13 − 4 = ☐

0 + 2 = ☐

14 ÷ 4 = ☐ REMAINDER ☐

5 − 3 = ☐

1 + 2 = ☐

5 × 7 = ☐

18 ÷ 6 = ☐

9 + 7 = ☐

3 × 4 = ☐

1 + 0 = ☐

14 ÷ 7 = ☐

6 − 1 = ☐

$3 - 1 =$ ☐

$6 \div 2 =$ ☐

$7 \times 3 =$ ☐

$7 + 2 =$ ☐

$18 \div 2 =$ ☐

$7 - 4 =$ ☐

$4 \times 5 =$ ☐

$6 + 4 =$ ☐

$9 \times 3 =$ ☐

$9 \div 4 =$ ☐ REMAINDER ☐

$8 - 1 =$ ☐

$9 - 6 =$ ☐

$19 \div 4 =$ ☐ REMAINDER ☐

$7 + 8 =$ ☐

$4 + 1 =$ ☐

$9 \div 1 =$ ☐

$4 \times 1 =$ ☐

$10 - 8 =$ ☐

$15 \div 6 =$ ☐ REMAINDER ☐

$6 \times 9 =$ ☐

$9 + 9 =$ ☐

$12 \div 2 =$ ☐

$6 + 2 =$ ☐

$3 \times 5 =$ ☐

$8 \div 6 =$ ☐ REMAINDER ☐

$2 - 1 =$ ☐

$7 + 9 =$ ☐

$8 \times 7 =$ ☐

$7 - 5 =$ ☐

$1 \div 1 =$ ☐

$0 \times 9 =$ ☐

$7 + 1 =$ ☐

$12 - 6 =$ ☐

$7 \times 6 =$ ☐

$8 \times 6 =$ ☐

$4 \times 4 =$ ☐

$16 - 7 =$ ☐

$14 \div 9 =$ ☐ REMAINDER ☐

$5 + 6 =$ ☐

$16 - 9 =$ ☐

$13 \div 8 =$ ☐ REMAINDER ☐

$3 + 5 =$ ☐

$9 + 1 =$ ☐

$1 \times 2 =$ ☐

$14 - 5 =$ ☐

$12 \div 4 =$ ☐

$7 \times 2 =$ ☐

$12 - 5 =$ ☐

$6 \div 6 =$ ☐

$4 + 2 =$ ☐

End Time ☐ : ☐

Date ☐ M ☐ D

Start Time ☐ : ☐

$16 \div 9 =$ ☐ REMAINDER ☐

$8 + 5 =$ ☐

$14 - 6 =$ ☐

$8 \times 3 =$ ☐

$2 + 1 =$ ☐

$7 \div 2 =$ ☐ REMAINDER ☐

$9 \times 2 =$ ☐

$13 - 6 =$ ☐

$6 + 7 =$ ☐

$2 \times 3 =$ ☐

$5 - 4 =$ ☐

$7 + 5 =$ ☐

$5 \times 8 =$ ☐

$8 \div 5 =$ ☐ REMAINDER ☐

$14 - 8 =$ ☐

$16 \div 2 =$ ☐

$3 - 3 =$ ☐

$2 + 9 =$ ☐

$9 \div 9 =$ ☐

$6 \times 7 =$ ☐

$2 + 3 =$ ☐

$15 \div 2 =$ ☐ REMAINDER ☐

$9 \times 8 =$ ☐

$19 \div 2 =$ ☐ REMAINDER ☐

$11 - 8 =$ ☐

$8 \times 4 =$ ☐

$6 - 2 =$ ☐

$15 - 8 =$ ☐

Lawn ☐

Pillow ☐

Swimmere ☐

Script ☐

$8 - 3 =$ ☐

$7 + 0 =$ ☐

$1 \times 5 =$ ☐

$8 \div 1 =$ ☐

$3 + 1 =$ ☐

$8 - 7 =$ ☐

$1 + 1 =$ ☐

$2 \times 2 =$ ☐

$14 \div 5 =$ ☐ REMAINDER ☐

$8 \div 2 =$ ☐

$8 \times 5 =$ ☐

$16 - 8 =$ ☐

$4 + 9 =$ ☐

$9 - 1 =$ ☐

$18 \div 9 =$ ☐

$1 \times 6 =$ ☐

$5 + 3 =$ ☐

Block

Time Required ☐ : ☐

$9 \times 1 =$ ☐

$12 \div 2 =$ ☐

$8 - 1 =$ ☐

$5 \times 5 =$ ☐

$1 + 6 =$ ☐

$19 \div 9 =$ ☐ REMAINDER ☐

$3 \times 7 =$ ☐

$16 \div 8 =$ ☐

$5 + 9 =$ ☐

$3 \div 3 =$ ☐

$13 - 8 =$ ☐

$9 + 6 =$ ☐

$8 - 5 =$ ☐

$12 \div 3 =$ ☐

$10 - 7 =$ ☐

$7 \times 4 =$ ☐

$1 + 4 =$ ☐

$18 \div 3 =$ ☐

$3 \times 8 =$ ☐

$9 - 6 =$ ☐

$5 + 1 =$ ☐

$7 \div 6 =$ ☐ REMAINDER ☐

$3 \times 9 =$ ☐

$14 \div 3 =$ ☐ REMAINDER ☐

$2 \times 4 =$ ☐

$3 + 6 =$ ☐

$12 - 4 =$ ☐

$7 + 4 =$ ☐

$4 - 3 =$ ☐

$7 \times 9 =$ ☐

$8 \div 8 =$ ☐

$6 + 6 =$ ☐

$7 \times 8 =$ ☐

$11 - 5 =$ ☐

$8 + 3 =$ ☐

$9 \times 4 =$ ☐

$17 - 8 =$ ☐

$10 \div 8 =$ ☐ REMAINDER ☐

$4 + 4 =$ ☐

$7 - 6 =$ ☐

$5 + 8 =$ ☐

$13 \div 6 =$ ☐ REMAINDER ☐

$8 \times 8 =$ ☐

$10 - 6 =$ ☐

$14 \div 2 =$ ☐

$2 \times 7 =$ ☐

$6 \times 8 =$ ☐

$9 - 0 =$ ☐

$9 \div 7 =$ ☐ REMAINDER ☐

$2 + 4 =$ ☐

End Time ☐ : ☐

I. Counting Test

Measure the time required for you to count from 1 to 120 aloud as fast as you can.

☐ sec.

II. Word Memorization Test

Memorize as many words as you can **within two minutes**.

lawn	province	harbour	hedgehog	flowerbed	uproar
pillow	noon	base	cooking	builder	frog
swimmer	finish	nameplate	limps	fan	courage
script	leaves	argument	pillar	peace	door
block	cross	smoke	snail	willow	hope

Write out as many words as you can remember **in two minutes** on the back of this page. How many words can you remember?

Number of words memorized ☐ words

Word Memorization Test Answers

<table>
<tr><td></td><td></td><td></td></tr>
<tr><td></td><td></td><td></td></tr>
<tr><td></td><td></td><td></td></tr>
<tr><td></td><td></td><td></td></tr>
<tr><td></td><td></td><td></td></tr>
<tr><td></td><td></td><td></td></tr>
<tr><td></td><td></td><td></td></tr>
<tr><td></td><td></td><td></td></tr>
<tr><td></td><td></td><td></td></tr>
<tr><td></td><td></td><td></td></tr>
</table>

III. Stroop Test

Please take the **Stroop Test** for Week 10, located on page **xiii** of the Appendix.

Start Time ☐ : ☐

$19 \div 4 =$ ☐ REMAINDER ☐

$2 + 8 =$ ☐

$8 - 5 =$ ☐

$9 \times 4 =$ ☐

$4 \div 3 =$ ☐ REMAINDER ☐

$5 \times 3 =$ ☐

$1 + 2 =$ ☐

$3 + 5 =$ ☐

$15 \div 3 =$ ☐

$9 \times 2 =$ ☐

$15 \div 8 =$ ☐ REMAINDER ☐

$7 - 4 =$ ☐

$15 - 8 =$ ☐

$2 \times 9 =$ ☐

$11 - 7 =$ ☐

$9 + 1 =$ ☐

$4 + 3 =$ ☐

$9 - 3 =$ ☐

$3 \times 8 =$ ☐

$9 \div 9 =$ ☐

$17 - 8 =$ ☐

$1 + 1 =$ ☐

$9 \div 2 =$ ☐ REMAINDER ☐

$9 \times 5 =$ ☐

$2 + 6 =$ ☐

$13 - 4 =$ ☐

$4 \times 2 =$ ☐

$11 \div 9 =$ ☐ REMAINDER ☐

$4 + 6 =$ ☐

$3 + 7 =$ ☐

$2 \times 6 =$ ☐

$5 - 2 =$ ☐

$7 \div 6 =$ ☐ REMAINDER ☐

$16 \div 8 =$ ☐

$7 \times 8 =$ ☐

$15 - 9 =$ ☐

$6 \div 3 =$ ☐

$6 + 1 =$ ☐

$9 - 0 =$ ☐

$15 \div 5 =$ ☐

$8 + 8 =$ ☐

$7 \times 6 =$ ☐

$6 - 1 =$ ☐

$4 + 0 =$ ☐

$9 \div 3 =$ ☐

$4 + 7 =$ ☐

$15 - 6 =$ ☐

$4 \times 1 =$ ☐

$5 \times 2 =$ ☐

$7 - 7 =$ ☐

$2 \times 5 =$ [ ]

$9 - 4 =$ [ ]

$8 + 7 =$ [ ]

$6 \times 7 =$ [ ]

$19 \div 6 =$ [ ] REMAINDER [ ]

$9 + 2 =$ [ ]

$9 \times 7 =$ [ ]

$6 + 7 =$ [ ]

$10 - 2 =$ [ ]

$16 \div 5 =$ [ ] REMAINDER [ ]

$9 - 5 =$ [ ]

$18 \div 3 =$ [ ]

$6 \div 2 =$ [ ]

$3 + 9 =$ [ ]

$8 - 3 =$ [ ]

$1 \times 8 =$ [ ]

$10 \div 5 =$ [ ]

$12 \div 8 =$ [ ] REMAINDER [ ]

$4 + 5 =$ [ ]

$12 - 3 =$ [ ]

$7 \div 1 =$ [ ]

$5 \times 7 =$ [ ]

$2 + 1 =$ [ ]

$6 \times 2 =$ [ ]

$10 - 1 =$ [ ]

$1 \times 6 =$ [ ]

$16 \div 2 =$ [ ]

$7 + 4 =$ [ ]

$8 - 7 =$ [ ]

$15 \div 7 =$ [ ] REMAINDER [ ]

$13 - 6 =$ [ ]

$9 \times 9 =$ [ ]

$5 \times 6 =$ [ ]

$5 + 0 =$ [ ]

$3 + 3 =$ [ ]

$14 - 5 =$ [ ]

$7 + 7 =$ [ ]

$3 \times 9 =$ [ ]

$17 \div 7 =$ [ ] REMAINDER [ ]

$12 \div 2 =$ [ ]

$4 \times 5 =$ [ ]

$7 - 2 =$ [ ]

$3 \times 7 =$ [ ]

$11 - 8 =$ [ ]

$6 \div 6 =$ [ ]

$4 + 4 =$ [ ]

$3 \times 6 =$ [ ]

$2 - 1 =$ [ ]

$3 + 6 =$ [ ]

$12 \div 9 =$ [ ] REMAINDER [ ]

End Time [ ] : [ ]

Date ☐ M ☐ D

Start Time ☐ : ☐

$8 \times 5 =$ ☐

$17 \div 9 =$ ☐ REMAINDER ☐

$8 + 6 =$ ☐

$12 - 5 =$ ☐

$9 + 4 =$ ☐

$17 \div 4 =$ ☐ REMAINDER ☐

$9 \times 6 =$ ☐

$12 - 9 =$ ☐

$3 - 2 =$ ☐

$8 \div 1 =$ ☐

$7 + 1 =$ ☐

$6 - 3 =$ ☐

$8 \times 4 =$ ☐

$12 \div 4 =$ ☐

$4 \times 7 =$ ☐

$9 + 8 =$ ☐

$12 \div 3 =$ ☐

$5 - 4 =$ ☐

$9 \times 3 =$ ☐

$7 - 1 =$ ☐

$8 + 5 =$ ☐

$2 \times 4 =$ ☐

$5 - 3 =$ ☐

$16 \div 4 =$ ☐

$2 + 3 =$ ☐

$8 \times 1 =$ ☐

$14 \div 6 =$ ☐ REMAINDER ☐

$4 + 1 =$ ☐

$10 - 5 =$ ☐

$2 \times 7 =$ ☐

$8 - 8 =$ ☐

$8 + 1 =$ ☐

$2 \div 1 =$ ☐

$6 \times 5 =$ ☐

$3 \div 2 =$ ☐ REMAINDER ☐

$7 \times 1 =$ ☐

$9 + 5 =$ ☐

$11 - 3 =$ ☐

$8 + 2 =$ ☐

$17 - 9 =$ ☐

$5 \div 5 =$ ☐

$1 \times 3 =$ ☐

$2 + 5 =$ ☐

$10 - 9 =$ ☐

$14 \div 5 =$ ☐ REMAINDER ☐

$1 + 6 =$ ☐

$3 \times 1 =$ ☐

$8 - 2 =$ ☐

$1 + 8 =$ ☐

$17 \div 2 =$ ☐ REMAINDER ☐

$18 \div 4 =$ ☐ REMAINDER ☐ $4 - 3 =$ ☐ $11 - 6 =$ ☐

$4 + 8 =$ ☐ $8 \div 4 =$ ☐ $8 \times 2 =$ ☐

$13 - 8 =$ ☐ $6 \times 9 =$ ☐ $6 + 6 =$ ☐

$3 \times 2 =$ ☐ $5 + 5 =$ ☐ $18 \div 6 =$ ☐

$4 \div 4 =$ ☐ $4 \div 2 =$ ☐ $2 + 2 =$ ☐

$6 \times 1 =$ ☐ $7 - 5 =$ ☐ $7 - 3 =$ ☐

$16 \div 6 =$ ☐ REMAINDER ☐ $14 \div 7 =$ ☐ $9 \div 8 =$ ☐ REMAINDER ☐

$7 + 3 =$ ☐ $8 \times 9 =$ ☐ $6 \times 4 =$ ☐

$12 - 4 =$ ☐ $5 + 3 =$ ☐ $10 \div 4 =$ ☐ REMAINDER ☐

$3 - 1 =$ ☐ $18 \div 9 =$ ☐ $6 \times 3 =$ ☐

$9 \times 8 =$ ☐ $0 \times 3 =$ ☐ $12 - 8 =$ ☐

$5 \times 5 =$ ☐ $12 - 7 =$ ☐ $8 + 9 =$ ☐

$8 \div 7 =$ ☐ REMAINDER ☐ $5 + 4 =$ ☐ $9 - 4 =$ ☐

$13 - 9 =$ ☐ $8 \div 2 =$ ☐ $7 \times 4 =$ ☐

$8 \times 0 =$ ☐ $4 \times 4 =$ ☐ $17 \div 8 =$ ☐ REMAINDER ☐

$3 + 1 =$ ☐ $7 + 2 =$ ☐ $6 + 9 =$ ☐

$1 + 7 =$ ☐ $8 - 3 =$ ☐   End Time ☐ : ☐

Day 53

Start Time ☐ : ☐

$6 - 3 =$ ☐

$4 + 5 =$ ☐

$8 \times 7 =$ ☐

$12 \div 6 =$ ☐

$15 - 9 =$ ☐

$3 \times 5 =$ ☐

$16 \div 7 =$ ☐ REMAINDER ☐

$11 - 3 =$ ☐

$6 + 3 =$ ☐

$15 \div 6 =$ ☐ REMAINDER ☐

$7 + 3 =$ ☐

$3 \times 0 =$ ☐

$13 \div 5 =$ ☐ REMAINDER ☐

$12 - 3 =$ ☐

$1 + 8 =$ ☐

$5 \times 9 =$ ☐

$3 + 6 =$ ☐

$7 \times 3 =$ ☐

$8 \div 7 =$ ☐ REMAINDER ☐

$2 - 2 =$ ☐

$4 \div 4 =$ ☐

$5 \times 3 =$ ☐

$8 - 0 =$ ☐

$3 + 5 =$ ☐

$8 - 8 =$ ☐

$4 \times 5 =$ ☐

$5 + 9 =$ ☐

$18 \div 9 =$ ☐

$7 + 5 =$ ☐

$12 - 7 =$ ☐

$3 \times 2 =$ ☐

$7 + 6 =$ ☐

$13 \div 2 =$ ☐ REMAINDER ☐

$12 \div 2 =$ ☐

$10 - 4 =$ ☐

$3 \times 7 =$ ☐

$3 + 4 =$ ☐

$13 - 4 =$ ☐

$14 \div 6 =$ ☐ REMAINDER ☐

$6 - 6 =$ ☐

$15 \div 5 =$ ☐

$6 + 9 =$ ☐

$5 - 5 =$ ☐

$1 \times 5 =$ ☐

$4 \times 6 =$ ☐

$4 + 7 =$ ☐

$3 \times 3 =$ ☐

$5 + 1 =$ ☐

$5 \div 5 =$ ☐

$8 - 4 =$ ☐

$7 - 3 =$ ☐ $8 + 9 =$ ☐ $7 \times 7 =$ ☐

$10 \div 2 =$ ☐ $1 \times 1 =$ ☐ $8 \div 5 =$ ☐ REMAINDER ☐

$15 - 7 =$ ☐ $3 - 2 =$ ☐ $11 - 8 =$ ☐

$5 + 2 =$ ☐ $14 \div 7 =$ ☐ $5 + 6 =$ ☐

$4 \times 3 =$ ☐ $1 \times 9 =$ ☐ $4 \times 9 =$ ☐

$12 \div 2 =$ ☐ $9 + 6 =$ ☐ $4 + 2 =$ ☐

$2 + 7 =$ ☐ $6 - 5 =$ ☐ $8 - 3 =$ ☐

$8 \times 4 =$ ☐ $12 \div 3 =$ ☐ $8 + 2 =$ ☐

$17 \div 4 =$ ☐ REMAINDER ☐ $3 + 0 =$ ☐ $6 \div 3 =$ ☐

$7 \times 5 =$ ☐ $8 \times 8 =$ ☐ $5 \times 8 =$ ☐

$2 + 4 =$ ☐ $2 \times 5 =$ ☐ $7 - 7 =$ ☐

$13 - 6 =$ ☐ $11 - 4 =$ ☐ $12 - 4 =$ ☐

$4 \div 2 =$ ☐ $8 + 5 =$ ☐ $16 \div 8 =$ ☐

$8 \times 9 =$ ☐ $11 \div 9 =$ ☐ REMAINDER ☐ $6 + 5 =$ ☐

$1 + 5 =$ ☐ $1 \times 6 =$ ☐ $3 \times 4 =$ ☐

$16 - 8 =$ ☐ $2 - 1 =$ ☐ $19 \div 6 =$ ☐ REMAINDER ☐

$13 \div 3 =$ ☐ REMAINDER ☐ $15 \div 7 =$ ☐ REMAINDER ☐   End Time ☐ : ☐

Date ☐ M ☐ D

Start Time ☐ : ☐

$8 \times 5 =$ ☐

$10 \div 6 =$ ☐ REMAINDER ☐

$3 + 2 =$ ☐

$12 - 9 =$ ☐

$5 - 3 =$ ☐

$2 + 5 =$ ☐

$8 \div 2 =$ ☐

$7 + 1 =$ ☐

$1 \times 3 =$ ☐

$4 \times 8 =$ ☐

$6 \div 6 =$ ☐

$4 \times 2 =$ ☐

$9 \div 9 =$ ☐

$6 \times 8 =$ ☐

$8 - 5 =$ ☐

$9 + 2 =$ ☐

$8 - 2 =$ ☐

$16 \div 4 =$ ☐

$4 + 8 =$ ☐

$5 \times 2 =$ ☐

$16 - 7 =$ ☐

$7 \times 8 =$ ☐

$10 \div 3 =$ ☐ REMAINDER ☐

$3 + 8 =$ ☐

$16 - 9 =$ ☐

$9 \div 3 =$ ☐

$6 + 6 =$ ☐

$7 - 5 =$ ☐

$1 + 3 =$ ☐

$11 - 5 =$ ☐

$9 \times 3 =$ ☐

$14 \div 5 =$ ☐ REMAINDER ☐

$9 + 0 =$ ☐

$9 + 8 =$ ☐

$6 \times 2 =$ ☐

$12 - 5 =$ ☐

$9 \times 8 =$ ☐

$2 + 6 =$ ☐

$8 \div 6 =$ ☐ REMAINDER ☐

$4 - 3 =$ ☐

$8 + 1 =$ ☐

$10 - 5 =$ ☐

$7 \times 2 =$ ☐

$18 \div 2 =$ ☐

$7 - 2 =$ ☐

$5 \times 7 =$ ☐

$16 \div 6 =$ ☐ REMAINDER ☐

$7 + 4 =$ ☐

$4 - 2 =$ ☐

$19 \div 7 =$ ☐ REMAINDER ☐

$12 \div 4 =$ ☐ $9 \div 5 =$ ☐ REMAINDER ☐ $6 + 1 =$ ☐

$17 - 9 =$ ☐ $8 \times 2 =$ ☐ $0 \times 2 =$ ☐

$4 + 4 =$ ☐ $9 - 7 =$ ☐ $8 - 1 =$ ☐

$5 + 3 =$ ☐ $1 \times 7 =$ ☐ $5 \times 1 =$ ☐

$2 \times 9 =$ ☐ $7 \div 4 =$ ☐ REMAINDER ☐ $5 + 4 =$ ☐

$16 \div 2 =$ ☐ $8 + 7 =$ ☐ $12 \div 9 =$ ☐ REMAINDER ☐

$12 \div 5 =$ ☐ REMAINDER ☐ $2 \times 3 =$ ☐ $7 \div 1 =$ ☐

$10 - 9 =$ ☐ $10 \div 5 =$ ☐ $3 \times 6 =$ ☐

$8 + 3 =$ ☐ $13 - 7 =$ ☐ $13 - 8 =$ ☐

$4 \times 7 =$ ☐ $3 + 9 =$ ☐ $4 + 9 =$ ☐

$9 - 4 =$ ☐ $5 \times 4 =$ ☐ $14 \div 2 =$ ☐

$7 \times 6 =$ ☐ $5 - 2 =$ ☐ $3 \div 1 =$ ☐

$3 - 3 =$ ☐ $1 + 7 =$ ☐ $6 - 5 =$ ☐

$6 \times 7 =$ ☐ $15 - 8 =$ ☐ $10 - 7 =$ ☐

$2 \times 6 =$ ☐ $17 \div 7 =$ ☐ REMAINDER ☐ $6 \times 6 =$ ☐

$19 \div 3 =$ ☐ REMAINDER ☐ $8 \div 1 =$ ☐ $9 + 4 =$ ☐

$2 + 2 =$ ☐ $1 + 9 =$ ☐   **End Time** ☐ : ☐

Start Time ☐ : ☐

$8 + 5 =$ ☐

$3 \times 5 =$ ☐

$2 \div 2 =$ ☐

$6 - 6 =$ ☐

$5 + 2 =$ ☐

$4 \times 2 =$ ☐

$10 - 4 =$ ☐

$17 \div 5 =$ ☐ REMAINDER ☐

$4 \div 2 =$ ☐

$2 \times 8 =$ ☐

$5 - 1 =$ ☐

$7 + 2 =$ ☐

$6 \div 6 =$ ☐

$4 + 9 =$ ☐

$9 - 6 =$ ☐

$3 \times 6 =$ ☐

$4 \times 8 =$ ☐

$10 \div 3 =$ ☐ REMAINDER ☐

$8 + 2 =$ ☐

$3 - 1 =$ ☐

$9 - 8 =$ ☐

$14 \div 7 =$ ☐

$8 \times 9 =$ ☐

$3 + 4 =$ ☐

$14 \div 6 =$ ☐ REMAINDER ☐

$4 + 5 =$ ☐

$14 - 6 =$ ☐

$2 + 5 =$ ☐

$8 \div 3 =$ ☐ REMAINDER ☐

$7 + 1 =$ ☐

$4 \times 6 =$ ☐

$9 \times 3 =$ ☐

$17 - 8 =$ ☐

$16 \div 6 =$ ☐ REMAINDER ☐

$6 \times 9 =$ ☐

$11 - 4 =$ ☐

$1 + 2 =$ ☐

$12 - 3 =$ ☐

$9 \times 4 =$ ☐

$4 - 2 =$ ☐

$5 + 6 =$ ☐

$6 - 3 =$ ☐

$7 \times 5 =$ ☐

$9 + 2 =$ ☐

$12 \div 3 =$ ☐

$10 \div 2 =$ ☐

$3 + 7 =$ ☐

$1 \times 2 =$ ☐

$11 - 3 =$ ☐

$5 \div 3 =$ ☐ REMAINDER ☐

$2 - 1 =$ ☐

$12 \div 2 =$ ☐

$2 + 3 =$ ☐

$7 \times 6 =$ ☐

$7 + 3 =$ ☐

$2 \div 1 =$ ☐

$9 - 9 =$ ☐

$5 \times 6 =$ ☐

$17 \div 3 =$ ☐ REMAINDER ☐

$2 \times 1 =$ ☐

$3 + 6 =$ ☐

$5 - 4 =$ ☐

$18 \div 3 =$ ☐

$6 \times 4 =$ ☐

$12 - 8 =$ ☐

$15 \div 4 =$ ☐ REMAINDER ☐

$9 + 8 =$ ☐

$16 \div 8 =$ ☐

$8 + 9 =$ ☐

$1 + 7 =$ ☐

$8 \times 1 =$ ☐

$6 + 7 =$ ☐

$7 \times 4 =$ ☐

$9 \div 3 =$ ☐

$5 \times 1 =$ ☐

$13 - 4 =$ ☐

$4 + 3 =$ ☐

$10 \div 5 =$ ☐

$6 \times 3 =$ ☐

$10 - 2 =$ ☐

$7 \times 3 =$ ☐

$15 - 7 =$ ☐

$16 \div 3 =$ ☐ REMAINDER ☐

$8 - 8 =$ ☐

$7 \div 6 =$ ☐ REMAINDER ☐

$7 \times 8 =$ ☐

$12 \div 4 =$ ☐

$2 + 1 =$ ☐

$12 - 9 =$ ☐

$7 \times 0 =$ ☐

$4 - 3 =$ ☐

$9 + 7 =$ ☐

$13 \div 9 =$ ☐ REMAINDER ☐

$4 + 1 =$ ☐

$10 - 5 =$ ☐

$8 \times 2 =$ ☐

$3 \times 1 =$ ☐

$18 \div 4 =$ ☐ REMAINDER ☐

$9 - 2 =$ ☐

$8 + 7 =$ ☐

End Time ☐ : ☐

I. Counting Test

Measure the time required for you to count from 1 to 120 aloud as fast as you can.

☐ sec.

II. Word Memorization Test

Memorize as many words as you can **within two minutes**.

cake	ambition	decision	worm	light	yawn
teardrop	stream	tool	price	head	midnight
hill	ice	seat	firefly	curtsey	outline
wool	identity	leftovers	catfish	badger	morning
sandpit	behind	seaweed	field	swallow	mansion

Write out as many words as you can remember **in two minutes** on the back of this page. How many words can you remember?

Number of words memorized ☐ words

Word Memorization Test Answers

III. Stroop Test

Please take the **Stroop Test** for Week 11, located on page **xiv** of the Appendix.

Start Time ☐ : ☐

$3 \times 9 =$ ☐

$7 - 1 =$ ☐

$1 \times 7 =$ ☐

$14 \div 5 =$ ☐ REMAINDER ☐

$8 \times 5 =$ ☐

$6 - 4 =$ ☐

$0 + 2 =$ ☐

$17 \div 4 =$ ☐ REMAINDER ☐

$8 - 3 =$ ☐

$7 + 4 =$ ☐

$12 \div 2 =$ ☐

$9 \times 5 =$ ☐

$4 + 7 =$ ☐

$13 - 6 =$ ☐

$11 \div 8 =$ ☐ REMAINDER ☐

$9 + 9 =$ ☐

$2 + 8 =$ ☐

$4 \times 3 =$ ☐

$9 - 1 =$ ☐

$15 \div 5 =$ ☐

$4 \times 7 =$ ☐

$7 - 5 =$ ☐

$16 \div 4 =$ ☐

$5 + 1 =$ ☐

$2 \times 6 =$ ☐

$17 \div 2 =$ ☐ REMAINDER ☐

$3 + 8 =$ ☐

$6 - 5 =$ ☐

$12 - 4 =$ ☐

$12 \div 5 =$ ☐ REMAINDER ☐

$8 + 3 =$ ☐

$2 \times 7 =$ ☐

$1 + 6 =$ ☐

$8 \div 8 =$ ☐

$3 + 5 =$ ☐

$10 - 8 =$ ☐

$2 \times 3 =$ ☐

$6 \div 3 =$ ☐

$5 \times 8 =$ ☐

$8 \times 4 =$ ☐

$2 + 6 =$ ☐

$6 \times 5 =$ ☐

$11 - 7 =$ ☐

$4 \div 1 =$ ☐

$10 - 7 =$ ☐

$3 + 2 =$ ☐

$10 - 6 =$ ☐

$4 \div 3 =$ ☐ REMAINDER ☐

$9 - 5 =$ ☐

$1 + 4 =$ ☐

$14 - 5 =$ ☐

$3 + 1 =$ ☐

$5 \times 4 =$ ☐

$14 \div 2 =$ ☐

$1 + 5 =$ ☐

$11 \div 6 =$ ☐ REMAINDER ☐

$1 + 3 =$ ☐

$2 \times 9 =$ ☐

$9 - 3 =$ ☐

$3 \times 4 =$ ☐

$16 - 9 =$ ☐

$18 \div 9 =$ ☐

$3 \times 3 =$ ☐

$6 + 9 =$ ☐

$13 \div 5 =$ ☐ REMAINDER ☐

$12 - 6 =$ ☐

$6 \times 7 =$ ☐

$3 - 2 =$ ☐

$4 \times 4 =$ ☐

$11 \div 2 =$ ☐ REMAINDER ☐

$1 + 9 =$ ☐

$5 - 3 =$ ☐

$9 \div 1 =$ ☐

$3 \times 7 =$ ☐

$13 - 9 =$ ☐

$7 + 7 =$ ☐

$3 \times 8 =$ ☐

$15 \div 3 =$ ☐

$5 - 2 =$ ☐

$6 \div 2 =$ ☐

$7 \times 1 =$ ☐

$5 + 1 =$ ☐

$19 \div 3 =$ ☐ REMAINDER ☐

$5 + 7 =$ ☐

$6 \div 4 =$ ☐ REMAINDER ☐

$6 + 3 =$ ☐

$16 - 8 =$ ☐

$2 \times 5 =$ ☐

$7 \div 4 =$ ☐ REMAINDER ☐

$8 - 7 =$ ☐

$9 \times 2 =$ ☐

$4 + 8 =$ ☐

$9 \times 8 =$ ☐

$9 + 3 =$ ☐

$6 \div 1 =$ ☐

$10 - 9 =$ ☐

$1 \times 3 =$ ☐

$18 \div 6 =$ ☐

$9 - 2 =$ ☐

$4 + 4 =$ ☐

End Time ☐ : ☐

Day 57

Start Time ☐ : ☐

$16 \div 8 =$ ☐

$7 - 2 =$ ☐

$4 + 6 =$ ☐

$7 \times 3 =$ ☐

$11 - 4 =$ ☐

$10 - 2 =$ ☐

$7 + 3 =$ ☐

$6 \div 1 =$ ☐

$5 \times 7 =$ ☐

$6 + 4 =$ ☐

$11 \div 3 =$ ☐ REMAINDER ☐

$4 + 1 =$ ☐

$9 \times 9 =$ ☐

$2 \div 2 =$ ☐

$7 \times 4 =$ ☐

$8 - 3 =$ ☐

$4 - 0 =$ ☐

$9 \times 4 =$ ☐

$12 \div 5 =$ ☐ REMAINDER ☐

$10 - 6 =$ ☐

$2 \times 6 =$ ☐

$10 - 7 =$ ☐

$1 + 1 =$ ☐

$8 \times 0 =$ ☐

$8 + 1 =$ ☐

$6 \div 6 =$ ☐

$3 + 3 =$ ☐

$4 - 3 =$ ☐

$7 + 2 =$ ☐

$12 \div 2 =$ ☐

$8 - 4 =$ ☐

$6 \times 2 =$ ☐

$11 \div 7 =$ ☐ REMAINDER ☐

$3 - 1 =$ ☐

$9 + 7 =$ ☐

$8 \times 9 =$ ☐

$16 \div 7 =$ ☐ REMAINDER ☐

$7 - 1 =$ ☐

$5 + 7 =$ ☐

$14 \div 6 =$ ☐ REMAINDER ☐

$3 \times 7 =$ ☐

$8 \div 7 =$ ☐ REMAINDER ☐

$1 + 2 =$ ☐

$7 \times 7 =$ ☐

$10 - 5 =$ ☐

$4 + 9 =$ ☐

$6 \times 5 =$ ☐

$2 + 3 =$ ☐

$12 - 5 =$ ☐

$10 \div 5 =$ ☐

$3 + 4 =$ ☐

$3 \div 1 =$ ☐

$6 - 1 =$ ☐

$3 \times 4 =$ ☐

$6 - 5 =$ ☐

$8 + 3 =$ ☐

$11 \div 4 =$ ☐ REMAINDER ☐

$9 + 3 =$ ☐

$6 \times 1 =$ ☐

$18 \div 6 =$ ☐

$5 - 4 =$ ☐

$2 \times 7 =$ ☐

$8 \times 3 =$ ☐

$10 - 4 =$ ☐

$13 \div 7 =$ ☐ REMAINDER ☐

$6 \times 3 =$ ☐

$2 + 2 =$ ☐

$16 - 9 =$ ☐

$8 \times 8 =$ ☐

$1 + 5 =$ ☐

$9 \div 3 =$ ☐

$6 + 7 =$ ☐

$17 \div 6 =$ ☐ REMAINDER ☐

$8 - 8 =$ ☐

$4 - 1 =$ ☐

$9 \times 7 =$ ☐

$17 \div 3 =$ ☐ REMAINDER ☐

$3 \times 3 =$ ☐

$5 + 9 =$ ☐

$17 \div 7 =$ ☐ REMAINDER ☐

$9 \times 5 =$ ☐

$14 - 9 =$ ☐

$18 \div 2 =$ ☐

$7 + 9 =$ ☐

$5 + 3 =$ ☐

$2 \div 1 =$ ☐

$5 \times 3 =$ ☐

$10 - 1 =$ ☐

$9 \times 8 =$ ☐

$6 \times 7 =$ ☐

$2 - 1 =$ ☐

$17 \div 2 =$ ☐ REMAINDER ☐

$8 \times 2 =$ ☐

$4 + 8 =$ ☐

$16 \div 2 =$ ☐

$1 + 8 =$ ☐

$16 - 7 =$ ☐

$4 + 4 =$ ☐

$13 - 6 =$ ☐

$6 \div 2 =$ ☐

End Time ☐ : ☐

Day 58

Start Time ☐ : ☐

2 + 5 = ☐

17 ÷ 5 = ☐ REMAINDER ☐

13 − 7 = ☐

3 × 9 = ☐

1 + 7 = ☐

16 ÷ 4 = ☐

8 + 6 = ☐

8 − 1 = ☐

3 × 1 = ☐

2 × 5 = ☐

9 ÷ 4 = ☐ REMAINDER ☐

4 × 9 = ☐

11 − 5 = ☐

16 ÷ 3 = ☐ REMAINDER ☐

3 + 1 = ☐

14 − 8 = ☐

7 − 7 = ☐

4 ÷ 4 = ☐

10 ÷ 6 = ☐ REMAINDER ☐

13 − 5 = ☐

4 × 3 = ☐

5 + 8 = ☐

2 × 3 = ☐

8 − 6 = ☐

5 ÷ 1 = ☐

6 + 6 = ☐

6 × 8 = ☐

13 ÷ 2 = ☐ REMAINDER ☐

11 − 3 = ☐

8 × 4 = ☐

3 + 2 = ☐

8 − 5 = ☐

4 + 3 = ☐

8 + 7 = ☐

7 − 4 = ☐

12 ÷ 6 = ☐

5 × 8 = ☐

4 ÷ 2 = ☐

1 × 2 = ☐

7 + 1 = ☐

10 ÷ 7 = ☐ REMAINDER ☐

3 + 6 = ☐

9 + 2 = ☐

8 − 7 = ☐

4 − 4 = ☐

7 ÷ 7 = ☐

5 × 5 = ☐

12 − 3 = ☐

8 + 8 = ☐

4 × 4 = ☐

$4 \div 1 =$ [   ]

$1 - 1 =$ [   ]

$19 \div 5 =$ [   ] REMAINDER [   ]

$4 - 2 =$ [   ]

$6 \times 4 =$ [   ]

$1 + 3 =$ [   ]

$2 \times 4 =$ [   ]

$12 \div 3 =$ [   ]

$3 + 5 =$ [   ]

$2 + 4 =$ [   ]

$2 \times 8 =$ [   ]

$18 \div 8 =$ [   ] REMAINDER [   ]

$2 \times 1 =$ [   ]

$4 + 7 =$ [   ]

$3 - 3 =$ [   ]

$18 \div 5 =$ [   ] REMAINDER [   ]

$10 - 3 =$ [   ]

$12 \div 4 =$ [   ]

$0 \times 7 =$ [   ]

$4 + 2 =$ [   ]

$5 - 2 =$ [   ]

$2 \times 9 =$ [   ]

$7 + 4 =$ [   ]

$6 \div 3 =$ [   ]

$16 \div 6 =$ [   ] REMAINDER [   ]

$15 - 8 =$ [   ]

$4 \times 1 =$ [   ]

$11 - 6 =$ [   ]

$4 \times 6 =$ [   ]

$1 + 9 =$ [   ]

$9 - 6 =$ [   ]

$7 \times 5 =$ [   ]

$18 \div 3 =$ [   ]

$5 + 4 =$ [   ]

$2 + 1 =$ [   ]

$18 \div 9 =$ [   ]

$11 - 2 =$ [   ]

$9 \times 1 =$ [   ]

$11 \div 5 =$ [   ] REMAINDER [   ]

$12 - 8 =$ [   ]

$11 \div 9 =$ [   ] REMAINDER [   ]

$3 + 7 =$ [   ]

$6 - 5 =$ [   ]

$1 \times 4 =$ [   ]

$7 + 7 =$ [   ]

$3 \times 8 =$ [   ]

$15 \div 3 =$ [   ]

$17 - 9 =$ [   ]

$9 + 1 =$ [   ]

$1 \times 8 =$ [   ]

End Time [   ] : [   ]

Start Time ☐ : ☐

$4 + 1 =$ ☐

$5 - 4 =$ ☐

$6 \times 4 =$ ☐

$13 \div 2 =$ ☐ REMAINDER ☐

$18 - 9 =$ ☐

$9 \times 4 =$ ☐

$1 + 6 =$ ☐

$4 - 4 =$ ☐

$14 \div 3 =$ ☐ REMAINDER ☐

$4 \times 6 =$ ☐

$3 + 1 =$ ☐

$4 \div 2 =$ ☐

$5 + 2 =$ ☐

$18 \div 2 =$ ☐

$3 \times 1 =$ ☐

$4 - 3 =$ ☐

$7 - 6 =$ ☐

$3 \times 9 =$ ☐

$8 + 8 =$ ☐

$10 \div 7 =$ ☐ REMAINDER ☐

$14 - 8 =$ ☐

$8 \times 7 =$ ☐

$5 + 4 =$ ☐

$8 - 2 -$ ☐

$10 \div 4 =$ ☐ REMAINDER ☐

$16 - 8 =$ ☐

$7 \times 3 =$ ☐

$4 + 7 =$ ☐

$4 \div 1 =$ ☐

$5 + 8 =$ ☐

$2 \times 2 =$ ☐

$18 \div 3 =$ ☐

$1 + 2 =$ ☐

$9 + 9 =$ ☐

$12 \div 2 =$ ☐

$6 \times 7 =$ ☐

$11 - 4 =$ ☐

$19 \div 7 =$ ☐ REMAINDER ☐

$5 + 7 =$ ☐

$5 - 2 =$ ☐

$1 \times 7 =$ ☐

$18 \div 6 =$ ☐

$7 - 5 =$ ☐

$1 \times 1 =$ ☐

$2 + 2 =$ ☐

$13 - 4 =$ ☐

$11 \div 9 =$ ☐ REMAINDER ☐

$15 - 7 =$ ☐

$7 \times 7 =$ ☐

$5 + 9 =$ ☐

$16 \div 2 =$ ☐

$6 + 2 =$ ☐

$14 - 6 =$ ☐

$2 \times 5 =$ ☐

$0 \times 3 =$ ☐

$13 - 5 =$ ☐

$6 + 0 =$ ☐

$8 \div 2 =$ ☐

$7 \times 9 =$ ☐

$9 - 5 =$ ☐

$1 + 8 =$ ☐

$4 \times 9 =$ ☐

$2 \div 2 =$ ☐

$5 - 3 =$ ☐

$3 \times 6 =$ ☐

$5 \div 3 =$ ☐ REMAINDER ☐

$7 + 6 =$ ☐

$6 + 4 =$ ☐

$7 \div 6 =$ ☐ REMAINDER ☐

$3 - 1 =$ ☐

$8 \times 8 =$ ☐

$4 \times 2 =$ ☐

$19 \div 6 =$ ☐ REMAINDER ☐

$2 - 1 =$ ☐

$5 + 3 =$ ☐

$12 \div 6 =$ ☐

$8 \times 5 =$ ☐

$17 - 9 =$ ☐

$4 + 6 =$ ☐

$18 \div 5 =$ ☐ REMAINDER ☐

$6 \times 2 =$ ☐

$9 - 9 =$ ☐

$17 \div 4 =$ ☐ REMAINDER ☐

$7 + 1 =$ ☐

$11 - 3 =$ ☐

$1 \times 4 =$ ☐

$15 \div 3 =$ ☐

$8 + 5 =$ ☐

$7 + 2 =$ ☐

$3 \times 3 =$ ☐

$10 - 6 =$ ☐

$10 \div 5 =$ ☐

$8 - 3 =$ ☐

$5 \times 8 =$ ☐

$11 \div 3 =$ ☐ REMAINDER ☐

$8 + 9 =$ ☐

$13 - 7 =$ ☐

$4 + 8 =$ ☐

$7 \div 1 =$ ☐

$9 \times 2 =$ ☐

End Time ☐ : ☐

Start Time ☐ : ☐

$14 \div 2 =$ ☐

$4 + 2 =$ ☐

$6 - 2 =$ ☐

$7 \times 5 =$ ☐

$7 \times 4 =$ ☐

$14 \div 8 =$ ☐ REMAINDER ☐

$9 - 8 =$ ☐

$12 - 3 =$ ☐

$8 + 3 =$ ☐

$4 + 5 =$ ☐

$9 \div 9 =$ ☐

$3 \times 2 =$ ☐

$7 - 1 =$ ☐

$11 - 9 =$ ☐

$8 - 7 =$ ☐

$9 + 2 =$ ☐

$2 \div 1 =$ ☐

$11 \div 8 =$ ☐ REMAINDER ☐

$4 \times 8 =$ ☐

$11 - 7 =$ ☐

$2 + 7 =$ ☐

$5 + 1 =$ ☐

$6 \times 8 =$ ☐

$12 - 9 =$ ☐

$6 \div 2 =$ ☐

$6 + 3 =$ ☐

$3 \times 5 =$ ☐

$9 \times 7 =$ ☐

$7 - 4 =$ ☐

$9 + 3 =$ ☐

$2 + 5 =$ ☐

$5 \div 2 =$ ☐ REMAINDER ☐

$10 \div 2 =$ ☐

$6 \div 3 =$ ☐

$3 + 9 =$ ☐

$1 \times 9 =$ ☐

$12 - 4 =$ ☐

$4 \times 3 =$ ☐

$9 + 5 =$ ☐

$9 \div 2 =$ ☐ REMAINDER ☐

$1 + 5 =$ ☐

$13 - 6 =$ ☐

$7 + 4 =$ ☐

$8 \times 2 =$ ☐

$9 \times 8 =$ ☐

$7 - 3 =$ ☐

$6 - 4 =$ ☐

$7 \div 3 =$ ☐ REMAINDER ☐

$5 \times 9 =$ ☐

$16 \div 7 =$ ☐ REMAINDER ☐

$3 - 2 =$ ☐

$7 \div 4 =$ ☐ REMAINDER ☐

$8 \times 6 =$ ☐

$5 \times 4 =$ ☐

$1 + 9 =$ ☐

$10 - 5 =$ ☐

$2 + 4 =$ ☐

$8 \div 4 =$ ☐

$9 + 6 =$ ☐

$7 \times 6 =$ ☐

$6 \times 1 =$ ☐

$13 \div 3 =$ ☐ REMAINDER ☐

$11 - 2 =$ ☐

$2 + 3 =$ ☐

$11 - 6 =$ ☐

$9 \div 3 =$ ☐

$15 \div 5 =$ ☐

$6 \times 6 =$ ☐

$10 - 7 =$ ☐

$3 + 6 =$ ☐

$12 \div 2 =$ ☐

$8 \times 4 =$ ☐

$4 - 2 =$ ☐

$6 + 8 =$ ☐

$6 \div 6 =$ ☐

$2 \times 3 =$ ☐

$2 + 8 =$ ☐

$14 \div 6 =$ ☐ REMAINDER ☐

$4 + 3 =$ ☐

$15 - 9 =$ ☐

$9 - 3 =$ ☐

$8 \times 7 =$ ☐

$4 \times 4 =$ ☐

$9 \div 8 =$ ☐ REMAINDER ☐

$3 + 8 =$ ☐

$4 \times 5 =$ ☐

$6 - 5 =$ ☐

$9 \div 1 =$ ☐

$7 + 5 =$ ☐

$2 \times 4 =$ ☐

$2 - 0 =$ ☐

$6 \times 3 =$ ☐

$16 - 7 =$ ☐

$19 \div 4 =$ ☐ REMAINDER ☐

$10 \div 6 =$ ☐ REMAINDER ☐

$4 \times 7 =$ ☐

$6 \div 1 =$ ☐

$1 + 4 =$ ☐

$9 - 9 =$ ☐

$1 + 7 =$ ☐

End Time ☐ : ☐

I. Counting Test

Measure the time required for you to count from 1 to 120 aloud as fast as you can.

☐ sec.

II. Word Memorization Test

Memorize as many words as you can **within two minutes.**

sugar	drawbridge	language	fox	garden	ghost
justice	snowstorm	caviar	eyebrow	kettle	sandal
fresh	climber	numeral	green	vase	mice
lava	acoustics	convenience	monster	sorbet	clothing
echo	weather	labour	structure	family	firewood

Write out as many words as you can remember **in two minutes** on the back of this page. How many words can you remember?

Number of words memorized ☐ words

Word Memorization Test Answers

<table>
<tr><td></td><td></td><td></td></tr>
<tr><td></td><td></td><td></td></tr>
<tr><td></td><td></td><td></td></tr>
<tr><td></td><td></td><td></td></tr>
<tr><td></td><td></td><td></td></tr>
<tr><td></td><td></td><td></td></tr>
<tr><td></td><td></td><td></td></tr>
<tr><td></td><td></td><td></td></tr>
<tr><td></td><td></td><td></td></tr>
<tr><td></td><td></td><td></td></tr>
</table>

III. Stroop Test

Please take the **Stroop Test** for Week 12, located on page **xv** of the Appendix.

Answers

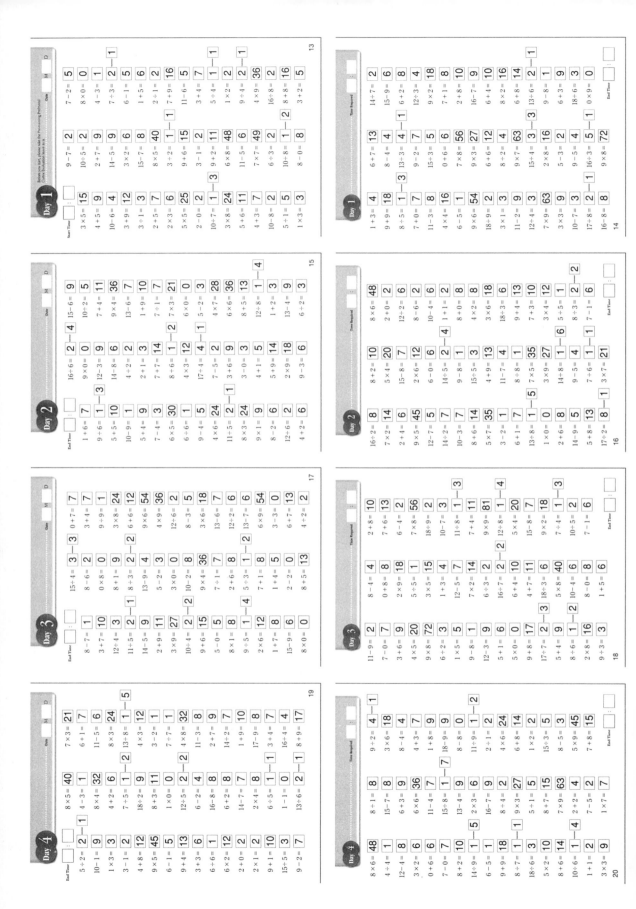

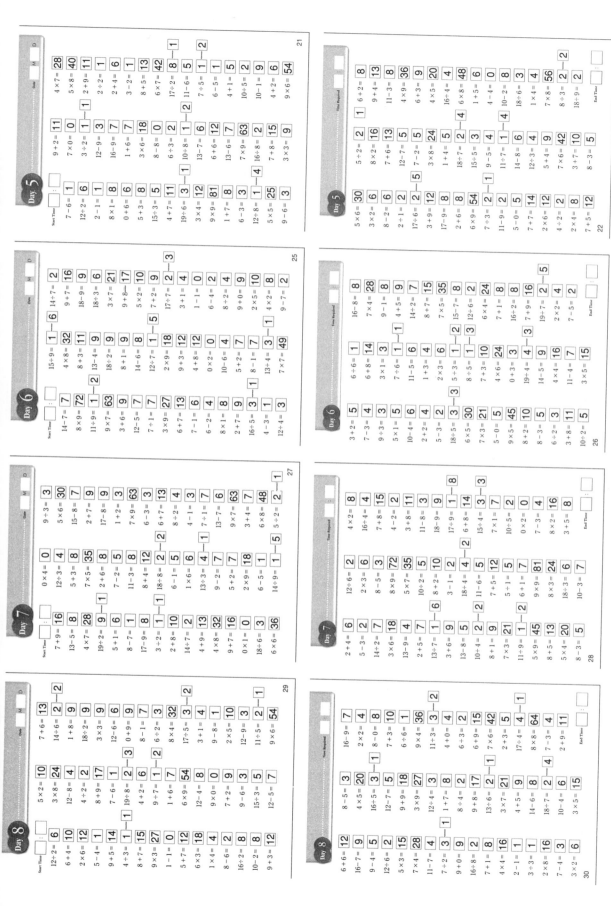

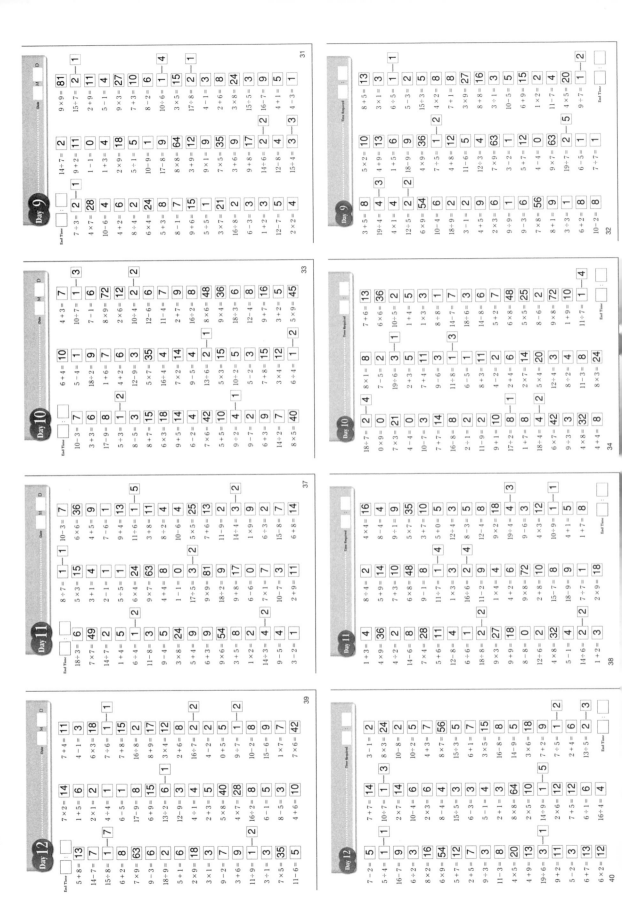

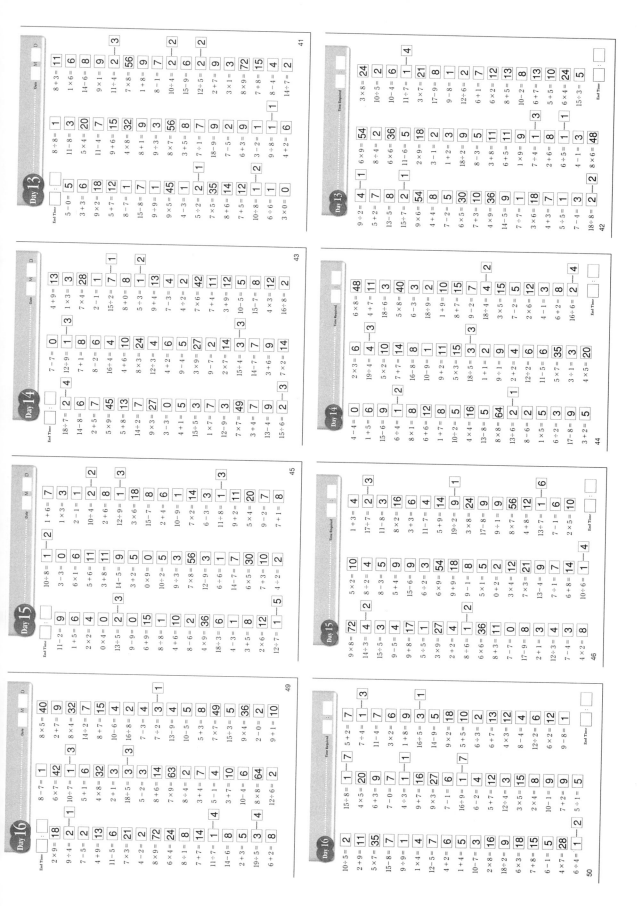

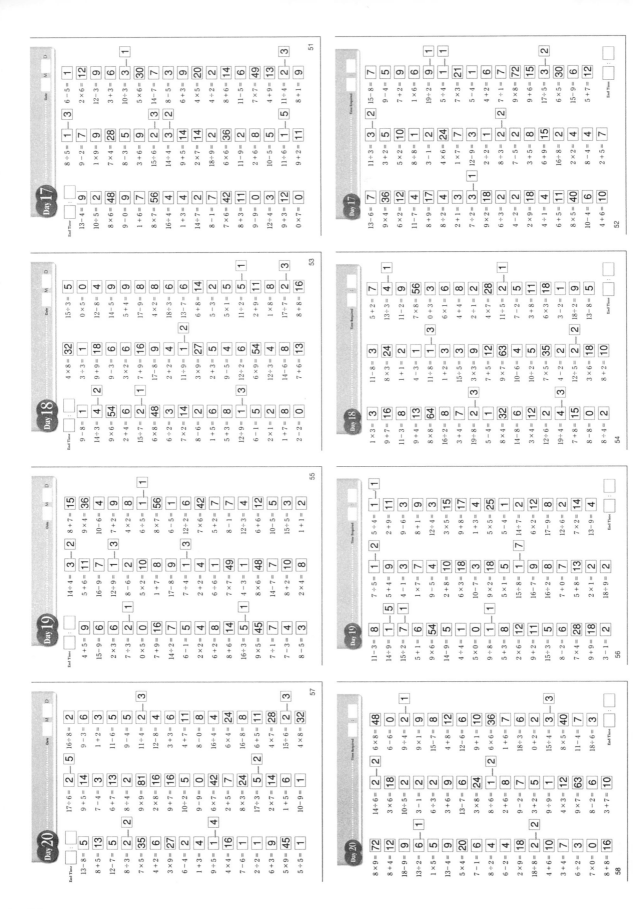

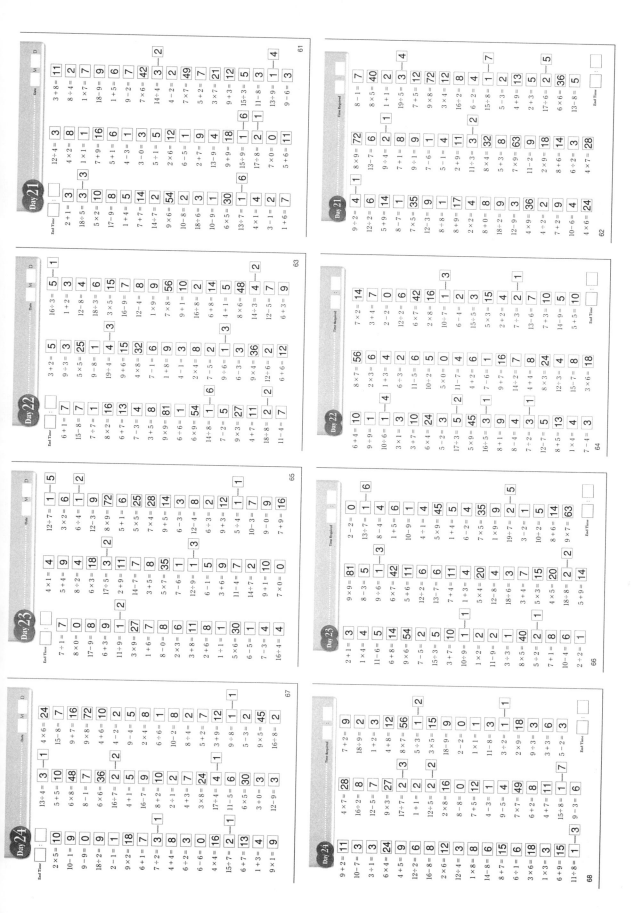

Day 21

Date M D

End Time :

2 + 1 = 3
18 ÷ 5 = 3 — 3
5 × 2 = 10
17 − 9 = 8
1 + 4 = 5
7 + 7 = 14
14 ÷ 7 = 2
9 × 6 = 54
10 − 8 = 2
18 ÷ 6 = 3
10 − 9 = 1
6 × 5 = 30
13 ÷ 7 = 1 — 6
4 × 1 = 4
3 − 1 = 2
1 + 6 = 7

12 ÷ 4 = 3
4 × 2 = 8
1 × 1 = 1
7 + 9 = 16
5 + 1 = 6
4 − 3 = 1
3 − 0 = 3
5 ÷ 1 = 5
2 × 6 = 12
6 − 5 = 1
2 + 7 = 9
13 − 9 = 4
9 + 9 = 18
15 ÷ 9 = 1 — 6
17 ÷ 8 = 2 — 1
7 × 0 = 0
5 + 6 = 11

3 + 8 = 11
8 ÷ 4 = 2
1 × 7 = 7
18 − 9 = 9
1 + 5 = 6
9 − 2 = 7
7 × 6 = 42
14 ÷ 4 = 3 — 2
4 − 2 = 2
7 × 7 = 49
5 + 2 = 7
3 × 7 = 21
9 + 3 = 12
15 ÷ 3 = 5
11 − 8 = 3
13 ÷ 9 = 1 — 4
9 − 6 = 3

61

Day 21

Time Required :

9 ÷ 2 = 4 — 1
12 ÷ 2 = 6
5 + 9 = 14
8 − 7 = 1
7 × 5 = 35
12 − 3 = 9
8 ÷ 8 = 1
8 + 9 = 17
2 × 2 = 4
8 + 0 = 8
18 ÷ 2 = 9
12 − 9 = 3
4 × 9 = 36
4 ÷ 2 = 2
7 + 2 = 9
10 − 6 = 4
4 × 6 = 24

8 × 9 = 72
13 − 7 = 6
9 ÷ 4 = 2 — 1
7 + 1 = 8
9 ÷ 1 = 9
7 − 6 = 1
5 − 1 = 4
2 + 9 = 11
11 ÷ 3 = 3 — 2
8 × 4 = 32
5 + 3 = 8
7 × 9 = 63
11 − 2 = 9
2 × 9 = 18
8 + 6 = 14
6 ÷ 2 = 3
4 × 7 = 28

8 − 1 = 7
8 × 5 = 40
1 + 1 = 2
19 ÷ 5 = 3 — 4
7 + 5 = 12
9 × 8 = 72
3 × 4 = 12
16 ÷ 2 = 8
6 − 2 = 4
15 ÷ 8 = 1 — 7
5 − 3 = 2
4 + 9 = 13
2 + 3 = 5
17 ÷ 6 = 2 — 5
6 × 6 = 36
13 − 8 = 5

End Time :

62

Day 22

Date M D

End Time :

6 + 1 = 7
15 − 8 = 7
7 ÷ 7 = 1
8 × 2 = 16
6 + 7 = 13
7 − 3 = 4
3 + 5 = 8
9 × 9 = 81
6 ÷ 6 = 1
6 × 9 = 54
14 ÷ 8 = 1 — 6
7 − 2 = 5
9 × 3 = 27
4 + 7 = 11
18 ÷ 8 = 2 — 2
11 − 4 = 7

3 + 2 = 5
9 ÷ 3 = 3
5 × 5 = 25
9 − 8 = 1
19 ÷ 4 = 4 — 3
9 + 6 = 15
4 × 8 = 32
7 − 1 = 6
1 + 8 = 9
4 − 1 = 3
2 × 4 = 8
7 − 5 = 2
9 ÷ 6 = 1 — 3
6 − 3 = 3
9 × 4 = 36
12 ÷ 6 = 2
6 + 6 = 12

16 ÷ 3 = 5 — 1
1 + 2 = 3
12 − 8 = 4
18 ÷ 3 = 6
3 × 5 = 15
16 − 9 = 7
12 − 4 = 8
1 × 9 = 9
7 × 8 = 56
9 + 1 = 10
16 ÷ 8 = 2
6 + 8 = 14
4 + 1 = 5
8 × 6 = 48
14 ÷ 3 = 4 — 2
12 − 5 = 7
6 + 3 = 9

63

Day 22

Time Required :

6 + 4 = 10
9 ÷ 9 = 1
10 ÷ 6 = 1 — 4
3 × 1 = 3
3 + 7 = 10
6 × 4 = 24
5 − 2 = 3
17 ÷ 3 = 5 — 2
5 × 9 = 45
16 ÷ 5 = 3 — 1
8 + 1 = 9
8 − 4 = 4
7 ÷ 2 = 3 — 1
12 − 7 = 5
8 + 5 = 13
1 × 4 = 4
7 − 4 = 3

8 × 7 = 56
2 × 3 = 6
1 + 3 = 4
6 ÷ 3 = 2
11 − 5 = 6
10 ÷ 2 = 5
5 × 0 = 0
11 − 7 = 4
4 + 2 = 6
7 − 6 = 1
9 + 7 = 16
14 ÷ 2 = 7
4 + 4 = 8
8 × 3 = 24
12 ÷ 3 = 4
15 − 7 = 8
3 × 6 = 18

7 × 2 = 14
3 + 4 = 7
2 − 2 = 0
12 ÷ 2 = 6
6 × 7 = 42
2 × 8 = 16
10 ÷ 7 = 1 — 3
6 − 4 = 2
15 ÷ 5 = 3
5 × 3 = 15
2 + 2 = 4
7 ÷ 3 = 2 — 1
13 − 6 = 7
7 + 3 = 10
14 − 9 = 5
5 + 5 = 10

End Time :

64

Day 23

Date M D

End Time :

7 ÷ 1 = 7
8 × 0 = 0
17 − 9 = 8
6 + 3 = 9
11 ÷ 9 = 1 — 2
3 × 9 = 27
1 + 6 = 7
8 − 0 = 8
2 × 3 = 6
3 + 8 = 11
2 + 6 = 8
1 ÷ 1 = 1
5 × 6 = 30
6 − 5 = 1
7 − 3 = 4
16 ÷ 4 = 4

4 × 1 = 4
5 + 4 = 9
8 ÷ 2 = 4
6 × 3 = 18
17 ÷ 5 = 3 — 2
2 + 9 = 11
14 − 7 = 7
3 + 5 = 8
5 × 7 = 35
7 − 6 = 1
12 ÷ 9 = 1 — 3
6 − 1 = 5
3 + 6 = 9
11 − 4 = 7
14 ÷ 7 = 2
9 + 1 = 10
7 × 0 = 0

12 ÷ 7 = 1 — 5
3 × 2 = 6
6 ÷ 4 = 1 — 2
12 − 3 = 9
8 × 9 = 72
5 + 1 = 6
5 × 5 = 25
7 × 4 = 28
9 + 5 = 14
6 − 3 = 3
12 − 4 = 8
6 ÷ 3 = 2
9 + 3 = 12
5 ÷ 4 = 1 — 1
10 − 3 = 7
9 − 0 = 9
7 + 9 = 16

65

Day 23

Time Required :

2 + 1 = 3
1 × 4 = 4
11 − 6 = 5
6 + 8 = 14
9 × 6 = 54
7 − 5 = 2
15 ÷ 3 = 5
3 + 7 = 10
10 ÷ 9 = 1 — 1
1 × 2 = 2
11 − 9 = 2
3 ÷ 3 = 1
8 × 5 = 40
5 ÷ 2 = 2 — 1
7 + 1 = 8
10 − 4 = 6
2 ÷ 2 = 1

9 × 9 = 81
8 − 3 = 5
9 ÷ 6 = 1 — 3
6 × 7 = 42
5 + 6 = 11
12 ÷ 2 = 6
13 − 7 = 6
7 + 4 = 11
1 + 3 = 4
5 × 4 = 20
12 − 8 = 4
18 ÷ 6 = 3
3 + 4 = 7
5 × 3 = 15
4 × 5 = 20
18 ÷ 8 = 2 — 2
5 + 9 = 14

2 − 2 = 0
13 ÷ 7 = 1 — 6
8 − 4 = 4
1 + 5 = 6
10 − 9 = 1
4 ÷ 1 = 4
5 × 9 = 45
1 + 4 = 5
6 − 2 = 4
7 × 5 = 35
1 × 9 = 9
19 ÷ 7 = 2 — 5
3 − 2 = 1
10 ÷ 2 = 5
8 + 6 = 14
9 × 7 = 63

End Time :

66

Day 24

Date M D

End Time :

2 × 5 = 10
10 − 1 = 9
9 − 9 = 0
18 ÷ 2 = 9
2 − 1 = 1
9 × 2 = 18
6 + 1 = 7
7 ÷ 2 = 3 — 1
4 + 4 = 8
6 ÷ 2 = 3
6 − 6 = 0
4 × 4 = 16
15 ÷ 7 = 2 — 1
6 + 7 = 13
1 + 3 = 4
9 × 1 = 9

13 ÷ 4 = 3 — 1
5 + 5 = 10
6 × 8 = 48
8 − 1 = 7
6 × 6 = 36
16 ÷ 7 = 2 — 2
4 + 1 = 5
16 − 7 = 9
8 + 2 = 10
2 ÷ 1 = 2
4 + 3 = 7
3 × 8 = 24
17 ÷ 4 = 4 — 1
11 − 5 = 6
6 × 5 = 30
3 + 0 = 3
12 − 9 = 3

4 × 6 = 24
15 − 8 = 7
9 + 7 = 16
9 × 8 = 72
4 + 6 = 10
4 − 2 = 2
9 − 4 = 5
2 × 4 = 8
6 ÷ 6 = 1
10 − 2 = 8
8 ÷ 4 = 2
5 + 2 = 7
3 + 9 = 12
9 ÷ 8 = 1 — 1
5 − 3 = 2
9 × 5 = 45
16 ÷ 8 = 2

67

Day 24

Time Required :

9 + 2 = 11
10 − 7 = 3
3 ÷ 1 = 3
6 × 4 = 24
4 + 5 = 9
12 ÷ 2 = 6
16 − 8 = 8
2 × 6 = 12
12 ÷ 4 = 3
1 × 8 = 8
14 − 8 = 6
8 + 7 = 15
6 ÷ 1 = 6
3 × 6 = 18
1 × 3 = 3
6 + 9 = 15
11 ÷ 8 = 1 — 3

4 × 7 = 28
16 ÷ 2 = 8
12 − 5 = 7
9 × 3 = 27
17 ÷ 7 = 2 — 3
1 + 1 = 2
12 ÷ 5 = 2 — 2
2 × 8 = 16
8 − 8 = 0
7 + 5 = 12
4 − 3 = 1
9 − 5 = 4
7 × 7 = 49
6 + 2 = 8
4 + 7 = 11
15 ÷ 8 = 1 — 7
9 − 3 = 6

7 + 2 = 9
18 ÷ 9 = 2
1 + 2 = 3
4 + 8 = 12
8 × 7 = 56
5 ÷ 3 = 1 — 2
3 × 5 = 15
18 − 9 = 9
2 − 2 = 0
1 × 1 = 1
11 − 8 = 3
3 ÷ 2 = 1 — 1
2 × 9 = 18
9 ÷ 3 = 3
3 + 3 = 6
5 − 2 = 3

End Time :

68

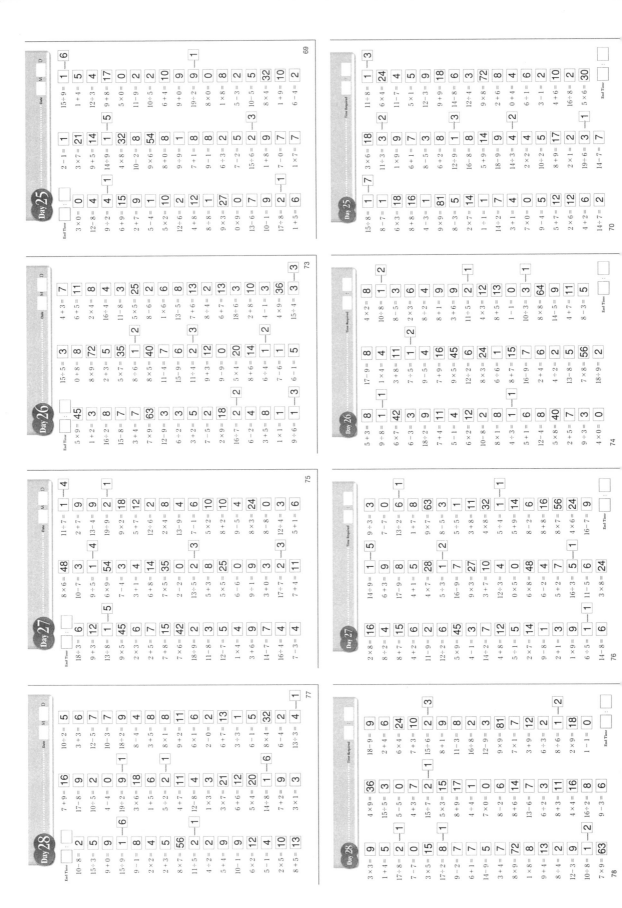

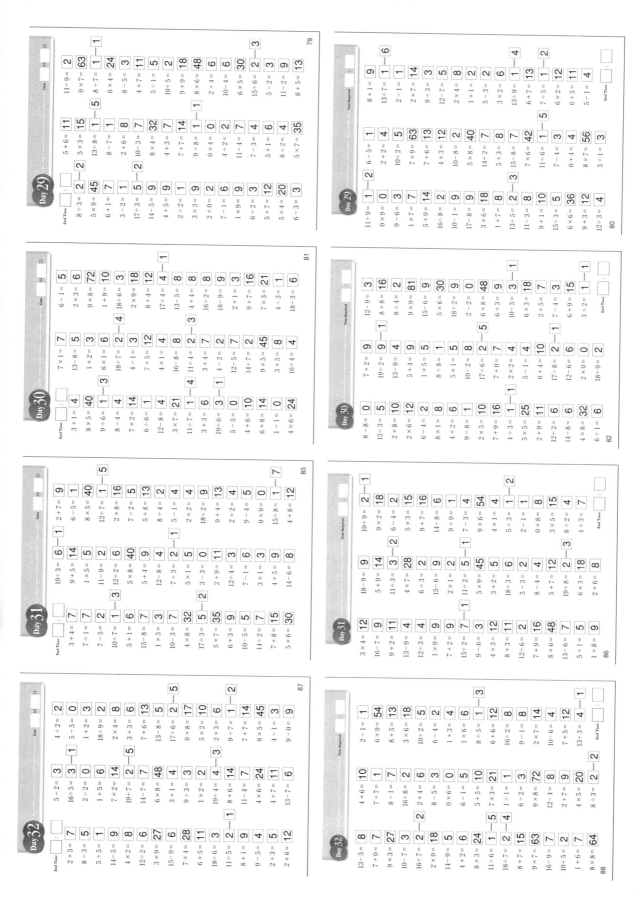

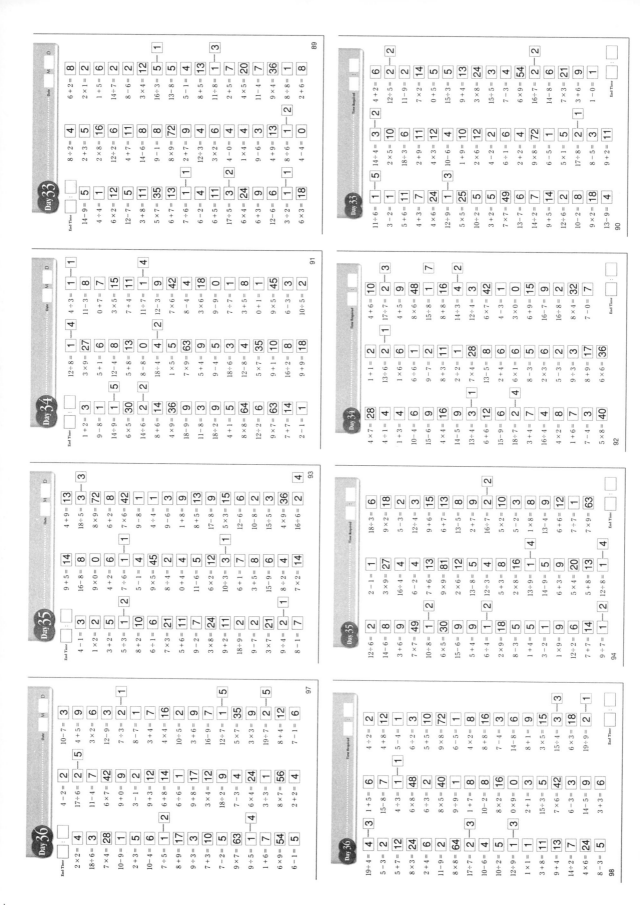

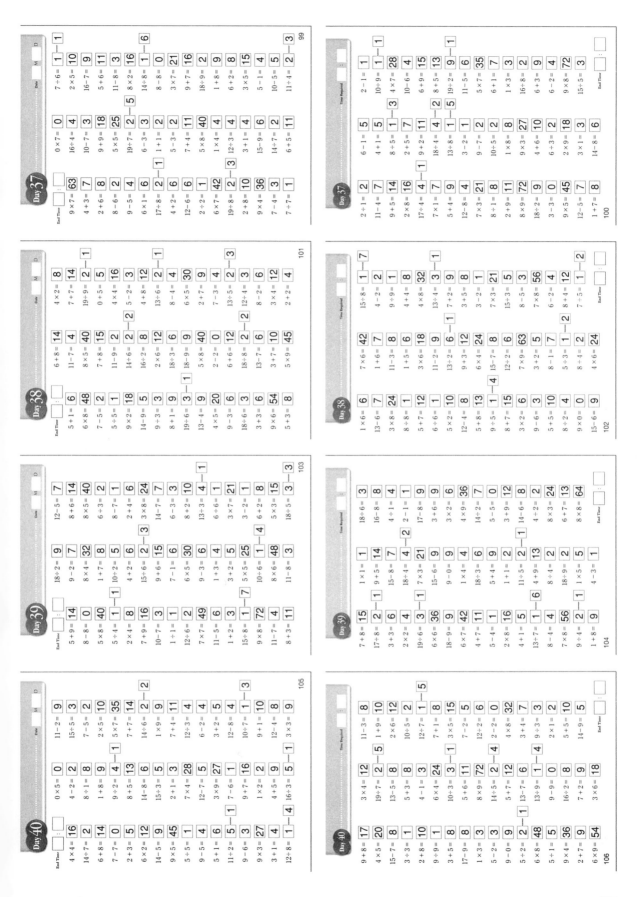

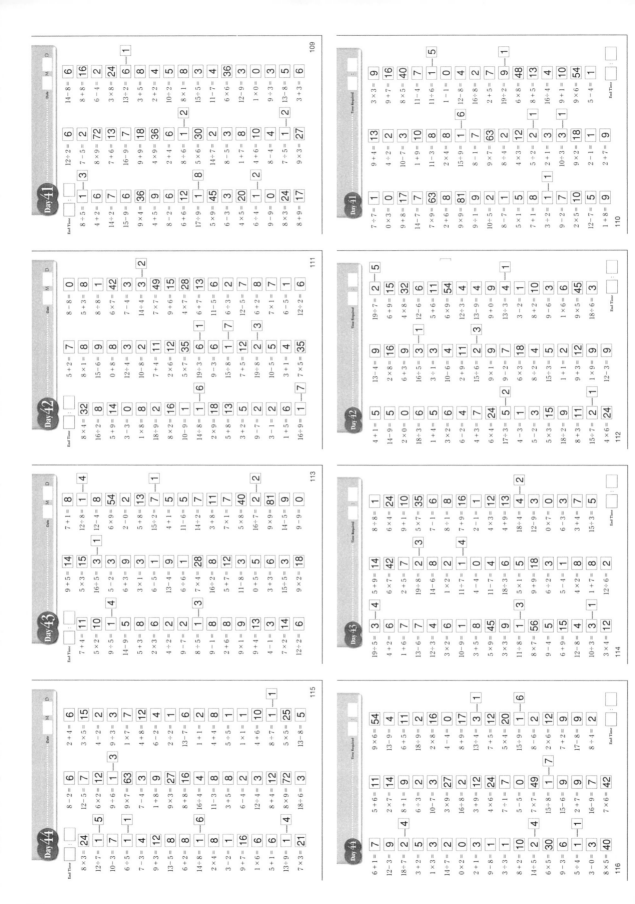

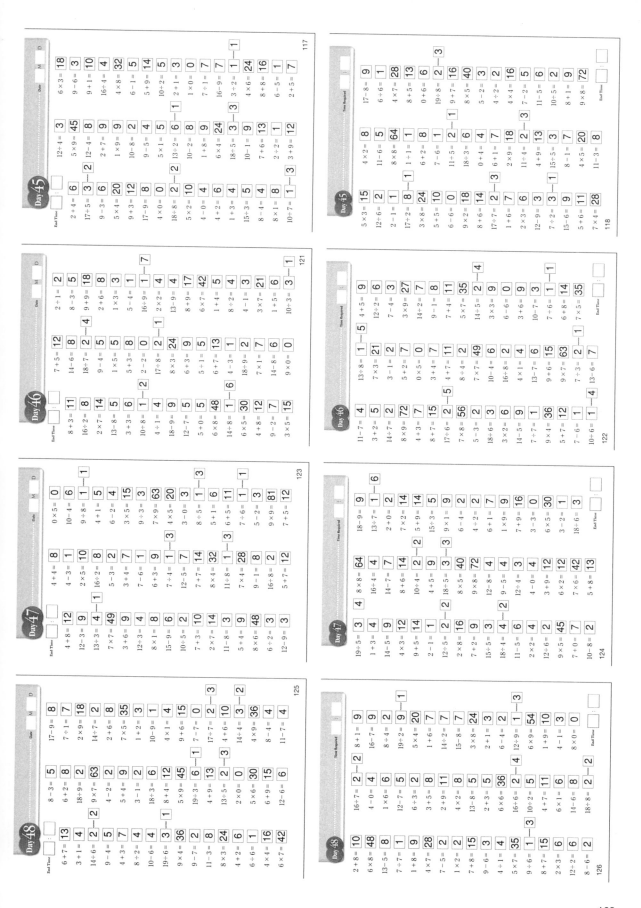

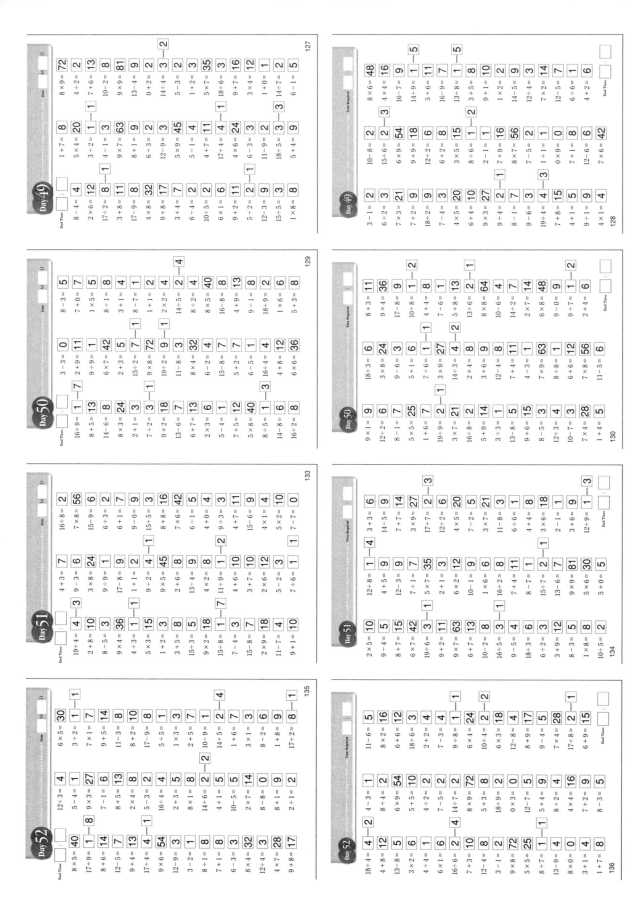

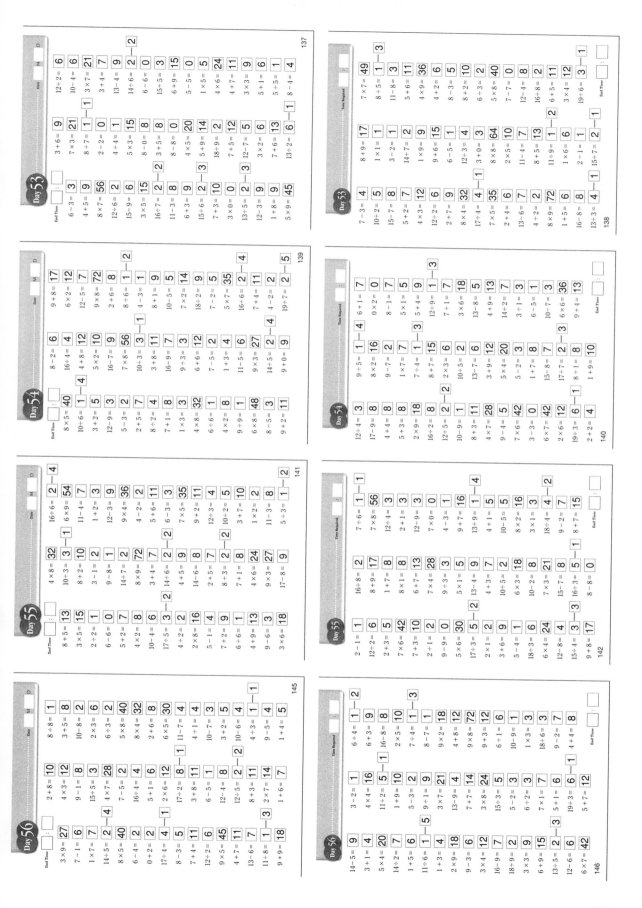

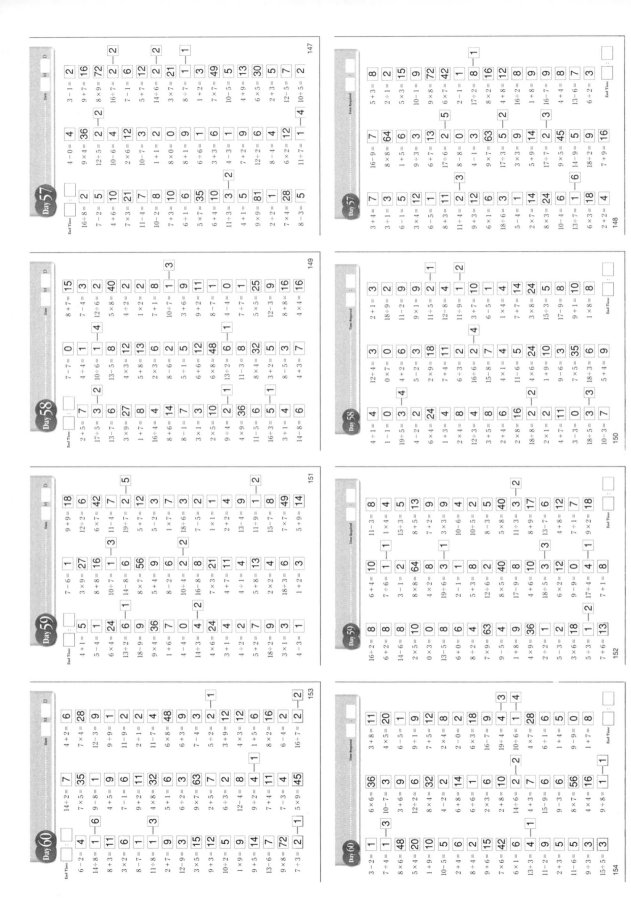

Pre-training Prefrontal Cortex Evaluation········ ☐ M ☐ D

Please evaluate your brain function before beginning the training.

I. Counting Test

Measure the time required for you to count from 1 to 120 aloud as fast as you can.

☐ sec.

II. Word Memorization Test

Memorize as many words as you can **within two minutes**.

secret	daughter	progress	rule	dust	children
clam	logic	chestnut	forest	wrestler	letter
mist	calm	present	blanket	speed	surface
fairytale	argument	easel	oxygen	cousin	name
midnight	cloth	bonfire	railway	splash	suburb

Write out as many words as you can remember **in two minutes** on the back of this page. How many words can you remember?

Number of words memorized ☐ words

Word Memorization Test Answers

Stroop Test (Pre-training)

This test should be taken only once. Before you take the test, please get used to the procedure by using the practice chart below. Name the colors the words are printed in aloud as fast as you can. If you make a mistake, name the color again correctly.

(Example : Red should be "**blue**." Red should be "**green**." Red should be "**red**.")

[Practice Chart]

Green   Red   Yellow   Green   Blue

Were you able to name the colors correctly (Blue, yellow, red, green and then yellow)?

Now let's proceed to the test. Fill in the "Start Time" box below and start naming the colors of the words printed on the chart. After you finish the test, fill in the "End Time" box and calculate the time required.

Start Time [ ] : [ ]

Red	Yellow	Green	Blue	Red
Red	Red	Yellow	Green	Blue
Blue	Yellow	Red	Blue	Green
Green	Green	Red	Blue	Yellow
Yellow	Blue	Green	Red	Yellow
Blue	Yellow	Red	Blue	Green
Green	Blue	Yellow	Red	Red
Yellow	Blue	Blue	Yellow	Blue
Green	Green	Red	Yellow	Blue
Red	Green	Green	Yellow	Blue

End Time [ ] : [ ]   **Time Required** [ ] : [ ]

Stroop Test (Week 1)

This test should be taken only once. Before you take the test, please get used to the procedure by using the practice chart below. Name the colors the words are printed in aloud as fast as you can. If you make a mistake, name the color again correctly.

(**Example :** Red should be "**blue**." Red should be "**green**." Red should be "**red**.")

[Practice Chart]

Green	Red	Yellow	Green	Blue

Were you able to name the colors correctly (Blue, yellow, red, green and then yellow)?

Now let's proceed to the test. Fill in the "Start Time" box below and start naming the colors of the words printed on the chart. After you finish the test, fill in the "End Time" box and calculate the time required.

Start Time [ ] : [ ]

Yellow	Blue	Green	Red	Red
Green	Blue	Yellow	Red	Red
Yellow	Yellow	Blue	Blue	Blue
Blue	Green	Red	Blue	Yellow
Yellow	Blue	Green	Red	Yellow
Green	Red	Green	Blue	Yellow
Red	Red	Blue	Yellow	Green
Red	Yellow	Green	Green	Blue
Red	Green	Green	Yellow	Blue
Blue	Green	Red	Blue	Yellow

End Time [ ] : [ ]   **Time Required** [ ] : [ ]

Stroop Test (Week 2)

This test should be taken only once. Before you take the test, please get used to the procedure by using the practice chart below. Name the colors the words are printed in aloud as fast as you can. If you make a mistake, name the color again correctly.

(**Example** : Red should be "**blue**." Red should be "**green**." Red should be "**red**.")

[Practice Chart]

| Green | Red | Yellow | Green | Blue |

Were you able to name the colors correctly (Blue, yellow, red, green and then yellow)?

Now let's proceed to the test. Fill in the "Start Time" box below and start naming the colors of the words printed on the chart. After you finish the test, fill in the "End Time" box and calculate the time required.

Start Time [   ] : [   ]

Red	Blue	Red	Yellow	Green
Green	Red	Green	Yellow	Blue
Green	Blue	Green	Yellow	Red
Blue	Yellow	Yellow	Blue	Green
Red	Blue	Red	Yellow	Green
Green	Yellow	Blue	Red	Red
Green	Yellow	Red	Green	Blue
Blue	Green	Yellow	Red	Blue
Red	Green	Yellow	Yellow	Blue
Blue	Blue	Red	Yellow	Green

End Time [   ] : [   ]   **Time Required** [   ] : [   ]

v

Stroop Test (Week 3)

This test should be taken only once. Before you take the test, please get used to the procedure by using the practice chart below. Name the colors the words are printed in aloud as fast as you can. If you make a mistake, name the color again correctly.

(**Example :** Red should be "**blue**." Red should be "**green**." Red should be "**red**.")

[Practice Chart]

Green	Red	Yellow	Green	Blue

Were you able to name the colors correctly (Blue, yellow, red, green and then yellow)?

Now let's proceed to the test. Fill in the "Start Time" box below and start naming the colors of the words printed on the chart. After you finish the test, fill in the "End Time" box and calculate the time required.

Start Time ☐ : ☐

Red	Green	Green	Yellow	Blue
Red	Blue	Red	Yellow	Green
Yellow	Red	Red	Green	Blue
Yellow	Blue	Green	Red	Blue
Red	Blue	Red	Yellow	Green
Green	Yellow	Red	Green	Blue
Blue	Blue	Yellow	Yellow	Green
Green	Yellow	Red	Green	Blue
Red	Yellow	Blue	Green	Yellow
Blue	Red	Yellow	Blue	Green

End Time ☐ : ☐   **Time Required** ☐ : ☐

Stroop Test (Week 4)

This test should be taken only once. Before you take the test, please get used to the procedure by using the practice chart below. Name the colors the words are printed in aloud as fast as you can. If you make a mistake, name the color again correctly.

(**Example** : Red should be "**blue**." Red should be "**green**." Red should be "**red**.")

[Practice Chart]

| Green | Red | Yellow | Green | Blue |

Were you able to name the colors correctly (Blue, yellow, red, green and then yellow)?

Now let's proceed to the test. Fill in the "Start Time" box below and start naming the colors of the words printed on the chart. After you finish the test, fill in the "End Time" box and calculate the time required.

Start Time ☐ : ☐

Green	Yellow	Red	Green	Blue
Yellow	Green	Green	Blue	Red
Red	Red	Yellow	Green	Blue
Blue	Green	Yellow	Blue	Red
Yellow	Green	Red	Blue	Red
Red	Green	Green	Yellow	Blue
Yellow	Blue	Blue	Yellow	Green
Red	Blue	Green	Red	Yellow
Green	Yellow	Blue	Yellow	Red
Blue	Blue	Red	Yellow	Green

End Time ☐ : ☐   **Time Required** ☐ : ☐

Stroop Test (Week 5)

This test should be taken only once. Before you take the test, please get used to the procedure by using the practice chart below. Name the colors the words are printed in aloud as fast as you can. If you make a mistake, name the color again correctly.

(**Example :** Red should be "**blue**." Red should be "**green**." Red should be "**red**.")

[Practice Chart]

Green	Red	Yellow	Green	Blue

Were you able to name the colors correctly (Blue, yellow, red, green and then yellow)?

Now let's proceed to the test. Fill in the "Start Time" box below and start naming the colors of the words printed on the chart. After you finish the test, fill in the "End Time" box and calculate the time required.

Start Time ☐ : ☐

Blue	Green	Red	Yellow	Green
Yellow	Blue	Green	Yellow	Red
Green	Yellow	Blue	Blue	Red
Yellow	Blue	Green	Red	Blue
Yellow	Yellow	Blue	Blue	Green
Blue	Green	Red	Red	Yellow
Blue	Green	Yellow	Green	Red
Red	Blue	Red	Yellow	Green
Green	Green	Yellow	Red	Blue
Red	Yellow	Red	Blue	Green

End Time ☐ : ☐   **Time Required** ☐ : ☐

Stroop Test (Week 6)

This test should be taken only once. Before you take the test, please get used to the procedure by using the practice chart below. Name the colors the words are printed in aloud as fast as you can. If you make a mistake, name the color again correctly.

(**Example : Red** should be "**blue**." **Red** should be "**green**." **Red** should be "**red**.")

[Practice Chart]

Green	Red	Yellow	Green	Blue

Were you able to name the colors correctly (Blue, yellow, red, green and then yellow)?

Now let's proceed to the test. Fill in the "Start Time" box below and start naming the colors of the words printed on the chart. After you finish the test, fill in the "End Time" box and calculate the time required.

Start Time [   ] : [   ]

Red	Yellow	Blue	Green	Red
Blue	Red	Green	Yellow	Blue
Red	Blue	Yellow	Blue	Green
Blue	Yellow	Green	Red	Red
Red	Blue	Green	Green	Yellow
Green	Red	Green	Blue	Yellow
Yellow	Red	Yellow	Blue	Green
Green	Green	Yellow	Blue	Red
Blue	Green	Red	Yellow	Red
Yellow	Yellow	Blue	Blue	Green

End Time [   ] : [   ]   **Time Required** [   ] : [   ]

Stroop Test (Week 7)

This test should be taken only once. Before you take the test, please get used to the procedure by using the practice chart below. Name the colors the words are printed in aloud as fast as you can. If you make a mistake, name the color again correctly.

(**Example :** Red should be "**blue**." Red should be "**green**." Red should be "**red**.")

[Practice Chart]

Green	Red	Yellow	Green	Blue

Were you able to name the colors correctly (Blue, yellow, red, green and then yellow)?

Now let's proceed to the test. Fill in the "Start Time" box below and start naming the colors of the words printed on the chart. After you finish the test, fill in the "End Time" box and calculate the time required.

Start Time ☐ : ☐

Red	Blue	Red	Yellow	Green
Green	Yellow	Red	Green	Blue
Blue	Blue	Yellow	Yellow	Green
Yellow	Blue	Green	Blue	Red
Green	Yellow	Red	Green	Blue
Green	Green	Red	Yellow	Blue
Red	Red	Yellow	Blue	Green
Red	Red	Green	Yellow	Blue
Red	Green	Yellow	Yellow	Blue
Blue	Red	Yellow	Blue	Green

End Time ☐ : ☐   **Time Required** ☐ : ☐

Stroop Test (Week 8)

This test should be taken only once. Before you take the test, please get used to the procedure by using the practice chart below. Name the colors the words are printed in aloud as fast as you can. If you make a mistake, name the color again correctly.

(**Example : Red** should be "**blue**." **Red** should be "**green**." **Red** should be "**red**.")

[Practice Chart]

Green	Red	Yellow	Green	Blue

Were you able to name the colors correctly (Blue, yellow, red, green and then yellow)?

Now let's proceed to the test. Fill in the "Start Time" box below and start naming the colors of the words printed on the chart. After you finish the test, fill in the "End Time" box and calculate the time required.

Start Time ☐ : ☐

Yellow	Green	Red	Green	Blue
Red	Yellow	Blue	Red	Green
Red	Red	Blue	Green	Yellow
Red	Green	Yellow	Yellow	Blue
Red	Blue	Red	Yellow	Green
Green	Yellow	Red	Green	Blue
Blue	Blue	Yellow	Yellow	Blue
Yellow	Blue	Green	Red	Blue
Green	Yellow	Red	Green	Blue
Blue	Yellow	Red	Blue	Green

End Time ☐ : ☐   **Time Required** ☐ : ☐

Stroop Test (Week 9)

This test should be taken only once. Before you take the test, please get used to the procedure by using the practice chart below. Name the colors the words are printed in aloud as fast as you can. If you make a mistake, name the color again correctly.

(**Example : Red** should be "**blue**." **Red** should be "**green**." **Red** should be "**red**.")

[Practice Chart]

Green	Red	Yellow	Green	Blue

Were you able to name the colors correctly (Blue, yellow, red, green and then yellow)?

Now let's proceed to the test. Fill in the "Start Time" box below and start naming the colors of the words printed on the chart. After you finish the test, fill in the "End Time" box and calculate the time required.

Start Time ☐ : ☐

Green	Red	Green	Yellow	Blue
Yellow	Green	Blue	Yellow	Red
Red	Yellow	Blue	Red	Green
Yellow	Green	Red	Green	Blue
Blue	Red	Yellow	Blue	Green
Red	Blue	Red	Green	Yellow
Green	Yellow	Red	Blue	Green
Red	Red	Green	Blue	Yellow
Blue	Blue	Blue	Yellow	Yellow
Blue	Red	Green	Blue	Yellow

End Time ☐ : ☐   **Time Required** ☐ : ☐

Stroop Test (Week 10)

   This test should be taken only once. Before you take the test, please get used to the procedure by using the practice chart below. Name the colors the words are printed in aloud as fast as you can. If you make a mistake, name the color again correctly.

   (**Example :** Red should be "**blue**." Red should be "**green**." Red should be "**red**.")

[Practice Chart]

Green	Red	Yellow	Green	Blue

   Were you able to name the colors correctly (Blue, yellow, red, green and then yellow)?

   Now let's proceed to the test. Fill in the "Start Time" box below and start naming the colors of the words printed on the chart. After you finish the test, fill in the "End Time" box and calculate the time required.

Start Time [ ] : [ ]

Blue	Green	Blue	Red	Yellow
Yellow	Green	Green	Red	Blue
Green	Yellow	Red	Yellow	Blue
Red	Red	Yellow	Green	Blue
Yellow	Blue	Blue	Blue	Yellow
Red	Green	Yellow	Green	Blue
Green	Yellow	Red	Blue	Red
Blue	Green	Green	Yellow	Red
Red	Red	Green	Yellow	Blue
Blue	Red	Blue	Green	Yellow

End Time [ ] : [ ]   **Time Required** [ ] : [ ]

Stroop Test (Week 11)

This test should be taken only once. Before you take the test, please get used to the procedure by using the practice chart below. Name the colors the words are printed in aloud as fast as you can. If you make a mistake, name the color again correctly.

(**Example :** Red should be "**blue**." Red should be "**green**." Red should be "**red**.")

[Practice Chart]

Green	Red	Yellow	Green	Blue

Were you able to name the colors correctly (Blue, yellow, red, green and then yellow)?

Now let's proceed to the test. Fill in the "Start Time" box below and start naming the colors of the words printed on the chart. After you finish the test, fill in the "End Time" box and calculate the time required.

Start Time ☐ : ☐

Blue	Red	Blue	Green	Yellow
Red	Blue	Green	Yellow	Red
Yellow	Green	Blue	Red	Red
Green	Red	Green	Blue	Yellow
Blue	Yellow	Blue	Blue	Yellow
Blue	Green	Red	Yellow	Blue
Green	Red	Blue	Yellow	Green
Green	Red	Yellow	Blue	Yellow
Yellow	Blue	Red	Green	Red
Blue	Yellow	Red	Green	Green

End Time ☐ : ☐   **Time Required** ☐ : ☐

Stroop Test (Week 12)

This test should be taken only once. Before you take the test, please get used to the procedure by using the practice chart below. Name the colors the words are printed in aloud as fast as you can. If you make a mistake, name the color again correctly.

(**Example :** Red should be "**blue**." Red should be "**green**." Red should be "**red**.")

[Practice Chart]

Green	Red	Yellow	Green	Blue

Were you able to name the colors correctly (Blue, yellow, red, green and then yellow)?

Now let's proceed to the test. Fill in the "Start Time" box below and start naming the colors of the words printed on the chart. After you finish the test, fill in the "End Time" box and calculate the time required.

Start Time [ ] : [ ]

Blue	Green	Red	Yellow	Blue
Green	Red	Blue	Yellow	Green
Yellow	Red	Yellow	Green	Blue
Blue	Yellow	Blue	Blue	Yellow
Red	Yellow	Green	Green	Blue
Blue	Green	Yellow	Blue	Red
Red	Red	Blue	Green	Yellow
Yellow	Blue	Green	Red	Red
Green	Blue	Red	Yellow	Red
Red	Green	Green	Yellow	Blue

End Time [ ] : [ ]   **Time Required** [ ] : [ ]